Peter Wyatt Squire

The Pharmacopoeias of Twenty-Five of the London Hospitals

Fifth Edition

Peter Wyatt Squire

The Pharmacopoeias of Twenty-Five of the London Hospitals
Fifth Edition

ISBN/EAN: 9783337225445

Printed in Europe, USA, Canada, Australia, Japan

Cover: Foto ©berggeist007 / pixelio.de

More available books at **www.hansebooks.com**

THE PHARMACOPŒIAS
OF THE
LONDON HOSPITALS.

THE PHARMACOPŒIAS

OF TWENTY-FIVE OF THE

LONDON HOSPITALS.

ARRANGED IN GROUPS FOR COMPARISON
EXCEPT THE CHILDREN'S, DUBLIN, AND GERMAN,
WHICH ARE PLACED IN THE ADDENDA.

BY

PETER SQUIRE.

FIFTH EDITION.

REVISED
BY
PETER WYATT SQUIRE
AND
ALFRED HERBERT SQUIRE,

JOINTLY CHEMISTS IN ORDINARY ON THE ESTABLISHMENT OF THE QUEEN;
CHEMISTS IN ORDINARY TO H.R.H. THE PRINCE OF WALES.

LONDON:
J. & A. CHURCHILL,
11, NEW BURLINGTON STREET.
1885.

LONDON
PRINTED BY J. F. VIRTUE AND CO., LIMITED
CITY ROAD.

CONTENTS.

	Page
ACIDA	1
ÆTHER CHLORICUS	2
AQUÆ	2
ARGENTI NITRAS MITIOR	2
BALNEA	2
BUGINARIA	6
CAPSULÆ	7
CATAPLASMATA	8
CAUSTICA	9
CHARTA	11
COLLODIA	11
COLLUNARIA	11
COLLYRIA	12
CONFECTIONES VEL ELECTUARIA	14
DECOCTA	16
ELECTUARIA. See CONFECTIONES.	
EMPLASTRA	18
ENEMATA	19
ESSENTIA	23
EXTRACTA	23
FOMENTUM. See FOTUS.	
FOTUS	24
FUMI	24
FUMIGATIO	25
GARGARISMATA	26
GLYCERINA	33
GOSSYPIA MEDICATA	34
GUTTÆ	37
HAUSTUS	40

CONTENTS.

	Page
Hydrargyri Oleas	68
Hydrargyri Oleas c. Morphiâ	68
Infusa	68
Inhalationes	68
Injectiones	69
Injectiones Hypodermicæ	72
Insufflationes	75
Lapis Divinus	77
Lincti	77
Linimenta	81
Lintea Medicata	86
Liquores	86
Lohoch. *See* Lincti.	
Lotiones	88
Misturæ	104
Nebulæ	158
Nitre-Paper. *See* Charta.	
Oleum Carbolatum	161
Pasta	161
Pastilli	163
Peptonised Foods	165
Pessi	166
Phenol Iodatum	168
Pigmenta	168
Pilulæ	173
Potassa cum Calce	204
Potus	204
Pulveres	205
Sanguis Bovinus Exsiccatus	217
Sapones	217
Solutiones	217
Spiritus	218
Suppositoria	220
Syrupi	221
Tincturæ	224
Trochisci	226
Unguenta	229
Vapores—Inhalationes	244
Vaselina	252
Vina	252

	Page
PHARMACOPŒIA OF THE HOSPITAL FOR SICK CHILDREN	254
DIET TABLES	274
PHARMACOPŒIA OF THE GERMAN HOSPITAL . .	305
PHARMACOPŒIA OF THE MEATH HOSPITAL AND COUNTY DUBLIN INFIRMARY	311
PHARMACOPŒIA OF THE ADELAIDE HOSPITAL, DUBLIN	314

PREFACE

TO THE FIFTH EDITION.

Since the publication of the Fourth Edition we have to record the lamented death of the Author of this work. The present Edition has been revised by his two sons, who assisted him with the previous one.

Thirteen of the Hospitals have issued new editions in which many formulæ have been altered and new remedies introduced.

Much care and time have been spent on the revision of the work with the hope that it will meet with the same approval which has been shown to the former Editions.

P. W. SQUIRE.
A. H. SQUIRE.

413, OXFORD STREET,
 March, 1885.

PREFACE

TO THE FOURTH EDITION.

SINCE the Third Edition (published March, 1874) there have been no less than sixteen new editions of the several Pharmacopœias of the London Hospitals, which have delayed the publishing of this work, for some of them are as recent as April, 1879.

It was intended to add the Edinburgh and Dublin Hospital Pharmacopœias, but I am informed by Dr. MacLagan that in the Edinburgh Hospitals the students are taught to prescribe by the "British Pharmacopœia;" they have no special formulæ. The principal Hospitals of Dublin are the Meath and Adelaide. The Pharmacopœias of these and the German Hospital, kindly sent to me by Dr. Hermann Weber, as well as that for the Hospital for Sick Children, are placed at the end, with the Diet Tables. The work is enlarged by these additions, but the price is not increased.

THE AUTHOR.

May, 1879.

PREFACE

TO THE SECOND EDITION.

THE Author, whilst engaged on the preparation of the British Pharmacopœia in 1863, collected the several Hospital Pharmacopœias of London, and made an alphabetical classification of all the preparations in them for the use of his Committee; these were afterwards published, and have been out of print a long time.

Since the publication of the British Pharmacopœia, all the London Hospitals have published new Pharmacopœias, except two (Westminster and Skin). It has been thought a fitting opportunity to publish a Second Edition of "The Pharmacopœias of the London Hospitals," which now number seventeen (instead of thirteen on the former occasion); and as these have been revised with great care by the Physicians and Surgeons of each Hospital, all the formulæ have an importance, each one applicable for its special purpose. Such a collection cannot fail to be useful to the profession generally, affording examples of prescribing every remedy in the most appropriate form, and, at the same time, presenting to the practitioner the different views taken by the seventeen Hospitals here included, as to the doses and menstruum employed for administration.

277, OXFORD STREET,
March, 1869.

The following is the list of the Hospitals, with the dates of publication:—

		Published in
CHARING CROSS	West Strand	1884
CHEST, DISEASES OF (ROYAL)	City Road	1876
CHEST, DISEASES OF (CITY OF LONDON)	Victoria Park	1878
CONSUMPTION	Brompton	1881
FEVER	Liverpool Road	1864
GREAT NORTHERN	Caledonian Road	1882
GUY'S	Borough	1879
KING'S COLLEGE	Portugal Street	1869
LONDON	Mile End	1882
LONDON OPHTHALMIC (ROYAL)	Liverpool Street	1879
MIDDLESEX	Berners Street	1883
ROYAL FREE	Gray's Inn Road	1882
SAMARITAN	Seymour Street	1881
ST. BARTHOLOMEW'S	Smithfield	1882
ST. GEORGE'S	Hyde Park	1865
ST. MARY'S	Paddington	1878
ST. THOMAS'S	Westminster Bridge	1867
SKIN, BRITISH	Marlborough Street	1884
THROAT	Golden Square	1881
UNIVERSITY COLLEGE	Gower Street	1879
WESTMINSTER	Broad Sanctuary	1876
WESTMINSTER OPHTHALMIC (ROYAL)	King William St.	1874
WOMEN'S	Soho Square	1883

ADDENDA.

			Page
LONDON:—			
CHILDREN, HOSPITAL FOR	Gt. Ormond Street	1883	254
GERMAN HOSPITAL	Dalston		305

EDINBURGH has no Hospital Pharmacopœias.

DUBLIN:—

MEATH	311
ADELAIDE	314

THE
PHARMACOPŒIAS
OF THE
LONDON HOSPITALS.

ACIDA.

Acetum Ipecacuanhæ.
    Ipecacuanha 1 oz.; Diluted Acetic Acid sufficient to make 20 oz. *Throat.*

Acidum Carbolicum Liquefactum.
    Carbolic Acid Crystals 16 drms.; Rectified Spirit 1 drm.; dissolve by heat. *Royal Chest.*
    Carbolic Acid liquefied by heat 8 drms.; Glycerine 1 drm.; mix. *Women.*

Acidum Hydrobromicum.
    Bromide of Potassium 120 grs.; Tartaric Acid 153 grs.; Water 1 oz. Dissolve the Bromide and then the Acid in the water, and set the mixture aside until precipitation ceases, then decant the clear solution. Dose 20 to 60 mins. *London.*

Acidum Hydrobromicum Dilutum.
    Bromide of Potassium 65 grs.; Cold Water 1 oz.; Tartaric Acid 80 grs. Dissolve the Bromide in the water then add the Acid, stir well together, put in a cool place for six hours then decant or filter. *Throat.*

Acidum Sulphuricum Alcoholisatum.
    Sulphuric Acid 1 fl. drm.; Rectified Spirit 7 fl. drms.; Oil of Sage 4 mins. Pour drop by drop the acid into the spirit, and when cold add the Oil of Sage. *Throat.*

ÆTHER CHLORICUS.
Chloroform 1 drm. ; Rectified Spirit to 10 drms. Dose 5 to 20 mins. *Consumption.*

AQUÆ.
Aqua Cassiæ.
   Oil of Cassia 5 mins.; Rectified Spirit 30 mins.; water to 10 oz. *St. George's.*

Aqua Glycyrrhizæ.
   Extract of Liquorice 20 grs.; water to 1 oz. *London.*
   Extract of Liquorice ½ oz.; water 20 oz. *Guy's.*

Aqua Medicatæ Omnes.
   Essential Oil 3 mins.; Rectified Spirit 57 mins.; water 12 oz. *London.*
   Essential Oil 2½ mins.; Rectified Spirit 2 drms.; water to 20 oz. *St. Bartholomew's.*
   Essential Oil 6 mins.; Distilled Water 1 oz. Shake well at short intervals for one day, and filter. *British Skin.*

Aqua Medicatæ Pulverisatæ. *See* Nebulæ.

ARGENTI NITRAS MITIOR.
Nitrate of Silver 1; Nitrate of Potash 3. Fuse together and cast in moulds. *Guy's.*

BALNEA.
Balneum Acidi Nitro-Hydrochlorici.
   Nitric Acid 1 lb.; Hydrochloric Acid 1½ lbs.; warm water 30 gals. *Guy's.*
   Nitric Acid 15 fl. oz.; Hydrochloric Acid 30 fl. oz.; water at 98° F., 30 gals. *London.*

Balneum Acidum.
   Diluted Nitro-hydrochloric Acid 12 oz.; water, 95° to 105° F., 30 gals. *St. Mary's. University.*
   Diluted Nitro-hydrochloric Acid 10 oz.; Water 30 gals. *British Skin.*
   Nitric Acid 1½ fl. oz.; Hydrochloric Acid 1 fl. oz.; water 30 gals. *Royal Free.*

Balneum Algense.
Carragheen Moss 16 oz. ; water 30 gals. Wash the moss in cold water to remove impurities; boil it for ¼ hour in 3 gals. of water; strain while hot ; wash the marc with boiling water to make up 3 gals., and mix the product with the water of the bath. *British Skin.*

Balneum Alkalinum.
Carbonate of Potash 2 oz. ; warm water 25 gals. *Charing Cross.*
Carbonate of Potash 3 oz.; Carbonate of Soda 3 oz.; warm water 30 gals. *Middlesex.*
Bicarbonate of Soda 4 oz. ; water, 95° to 105° F., 30 gals. *St. Mary's. University.*
Crystal. Carbonate Soda 4 oz. ; hot water 30 gals. *British Skin. Guy's.*
Crystal. Carbonate Soda 4, 6, or 8 oz.; water at 90° F., 30 gals. *London.*
Carbonate of Soda Crystals 8 oz. ; water 30 gals. *Royal Free. St. Bartholomew's.*

Balneum Alkalinum Compositum.
Carbonate of Soda 6 oz. ; Borax 1½ oz.; water at 98° F., 30 gals. ; dissolve and add Prepared Starch 4 oz. *London.*

Balneum Avenæ Farinæ.
Oatmeal 4 lbs.; water at 98° F., 30 gals. *London.*

Balneum Bituminis Comp.
Compound Liquor of Bitumen 8 oz. ; water at 98° F., 30 gals. *London.*

Balneum Calidæ Aquæ.
London, 96° to 115° F., *Guy's,* 98° to 110° F.

Balneum Calidi Aëris.
Guy's, 120° to 150° F.

Balneum Calidum.
Water at 96° F., 30 gals. *St. Bartholomew's.*

Balneum Creasoti.
Creasote ¼ oz. ; Glycerine 2 oz. ; water, 98° F., 30 gals. *London.*
Creasote 2 oz. ; water 30 gals. *British Skin.*

Balneum Diuturnum.
: Water at 92° F., 30 gals. *British Skin.*

Balneum Frigidæ Aquæ.
: London 60° F., or reduced by means of ice to nearly 32° F. *Guy's* 60° F.

Balneum Furfuris.
: Wheaten Bran 4 lbs.; water 30 gals. Tie the bran in a muslin bag; macerate it with frequent agitation for 10 minutes in 10 gals. of water at 185° F.; then add more water to produce the requisite quantity at 95° to 105° F. *University. London* 98° F.
: Wheaten Bran 2 lbs.; water 30 gals. Boil the bran for ¼ hour in 2 gals. of water, strain and mix the solution with the water of the bath. *British Shin.*

Balneum Glutinis.
: Bran 5 lbs.; boiling water 2 gals.; water to 30 gals. *St. Bartholomew's.*
: Clarified Size 2 lbs.; water, 95° to 105° F., 30 gals. Dissolve. *University.*
: Size 6 lbs.; boiling water 1 gal.; water at 98° F. to 30 gals. *London.*
: Size 8 lbs.; boiling water 1 gal.; water to 30 gals. *Royal Free.*
: Patent Size 8 lbs.; water to 30 gals. *British Skin.*

Balneum Hydrargyri.
: Perchloride of Mercury 60 to 180 grs.; Chloride of Ammonium 1 to 3 drms.; water at 98° F., 30 gals. *London.*

Balneum Iodi Comp.
: Iodine ¼ oz.; Iodide of Potassium ½ oz.; water at 98° F., 30 gals. *London.*

Balneum Marinum, vel Sodii Chloridi.
: Sea Salt 10 lbs.; water 30 gals. *British Skin.*
: Bay Salt 8 lbs.; water, 98° F., 30 gals. *London.*

Balneum Potassæ Sulphuratæ v. Sulphuratum.
: Sulphurated Potash 4 oz.; water, 95° to 105° F., 30 gals. Dissolve. *British Skin. London* 98° F. *St. Mary's. University.*

Sulphurated Potash 5 oz.; water 30 gals. *Royal Free.*
Sulphurated Potash 6 oz.; warm water 30 gals. *Charing Cross.*
Sulphurated Potash 8 oz.; warm water 30 gals. *Middlesex.*
Sulphurated Potash 8 oz.; water 30 gals. *St. Bartholomew's.*

Balneum Saponatum.

White Castile Soap 1 lb.; water 30 gals. Dissolve the Soap in half a gallon of boiling water and mix the solution with the water of the bath. *British Skin.*

Balneum Sinapis.

Mustard 2 lbs.; water at 98° F., 30 gals. Add the Mustard gradually to a gallon of the water and allow to stand for 15 minutes; then pour the mixture into a coarse linen bag, and press out the liquid, which is to be stirred up with the bath. *London.*

Balneum Sodii Chloridi.

Bay Salt 8 lbs.; water at 98° F., 30 gals. *London.*

Balneum Sulphureum. *See* Balneum Potassæ Sulphuratæ.

Balneum Sulphuris Comp.

Precipitated Sulphur 4 oz.; Hyposulphite of Soda 1 oz.; Dil. Sulphuric Acid 1½ drm.; water at 98° F., 30 gals. *London.*

Balneum Tepidæ Aquæ.

St. Bartholomew's, 90° F. *Guy's, London*, 85° to 92° F.

Balneum Turcicum (Turkish Bath).

A hot dry air bath 100° to 200° F., with kneading or shampooing, and warm ablutions or frictions, ending with tepid or cold affusion. *London.*

Balneum Vaporis.

The vapour of hot water applied to the body or any part of it, by a suitable apparatus. *London.*

BUGINARIA.
(NASAL BOUGIES.)

Gelato-Glycerine.
: Refined Gelatine (by weight) 5 oz.; Glycerine (by weight) 6 oz.; water (by weight) 6 oz. Soak the Gelatine in the water for 12 hours with occasional stirring, add the glycerine, dissolve in a water-bath, and evaporate to produce 15 oz. by weight of the Gelato-glycerine. In making Bougies the Gelato-glycerine must be melted, the medicament added in the manner hereinafter described, and the substance poured into moulds of such a shape that each Bougie has a length of eight centimètres, and is of a tapering form, the diameter of the larger end being eight millimètres and that of the smaller extremity three millimètres. *Throat*.

Buginarium Acidi Carbolici.
: Carbolic Acid $\frac{1}{2}$ gr.; Gelato-glycerine 40 grs. To the Gelato-glycerine, melted in a water-bath, add the Carbolic Acid. Dissolve and pour into the mould. When solidified, remove for use. *Throat*.

Buginarium Bismuthi.
: Subnitrate of Bismuth 5 grs.; Glycerine 3 mins. Rub together, and add the mixture to Gelato-glycerine, melted in a water-bath, 40 grs. Mix and pour into the mould; when solidified, remove for use. *Throat*.

Buginarium Cupri Sulphatis.
: Sulphate of Copper, in powder, $\frac{1}{10}$ gr.; Gelato-glycerine 40 grs. To the Gelato-glycerine, melted in a water-bath, add the Sulphate of Copper. Dissolve and pour into the mould. When solidified, remove for use. *Throat*.

Buginarium Iodoformi.
: Iodoform, in fine powder, $\frac{1}{2}$ gr; Glycerine 1 min. Rub together, and add the mixture to Gelato-glycerine, melted in a water-bath, 40 grs. Mix and pour into the mould. When solidified, remove for use. *Throat*.

Buginarium Morphiæ.
: Acetate of Morphia $\frac{1}{10}$ gr.; Gelato-glycerine 40 grs. To the Gelato-glycerine, melted in a water-bath, add the Acetate of Morphia. Dissolve and pour into the mould. When solidified, remove for use. *Throat.*

Buginarium Pini Sylvestris.
: Oil of Scotch Pine Leaves, $\frac{1}{2}$ min.; Gelato-glycerine 40 grs. To the Gelato-glycerine, melted in a water-bath, add the Oil. Mix thoroughly and pour into the mould. When solidified, remove for use. *Throat.*

Buginarium Plumbi Acetatis.
: Acetate of Lead $\frac{1}{2}$ gr.; Glycerine 2 mins. Dissolve, and add to Gelato-glycerine, melted in a water-bath, 40 grs. Mix and pour into the mould. When solidified, remove for use. *Throat.*

Buginarium Thymol.
: Thymol $\frac{1}{10}$ gr.; Rectified Spirit $\frac{1}{2}$ min. Dissolve and add to Gelato-glycerine, melted in a water-bath, 40 grs. Mix thoroughly and pour into the mould. When solidified, remove for use. *Throat.*

Buginarium Zinci Sulphatis.
: Sulphate of Zinc, in powder, $\frac{1}{10}$ gr.; Gelato-glycerine 40 grs. To the Gelato glycerine, melted in a water-bath, add the Sulphate of Zinc. Dissolve and pour into the mould. When solidified, remove for use. *Throat.*

CAPSULÆ.

Capsulæ Olei Phosphorati.
: Each containing $\frac{1}{30}$ gr. of Phosphorus dissolved in Olive Oil. *British Skin.*
: Each containing $\frac{1}{60}$ or $\frac{1}{30}$ gr. of Phosphorus. *London.*

Capsulæ Olei Santali Flavi.
: Each contains 10 mins. of the oil. *London.*

CATAPLASMATA.

Cataplasma Acidi Carbolici.
Make a Linseed poultice, but substitute the Carbolic Acid lotion for one half of the water. *Fever.*

Cataplasma Carbonis.
Wood Charcoal ½ oz.; Linseed Meal 4 oz.; boiling water 10 oz. *Fever.*
Wood Charcoal 1 oz.; Linseed Meal 3 oz.; boiling water 10 oz. *London.*

Cataplasma Carotæ.
Carrots boiled until they are soft, and pulp them. *Guy's.*

Cataplasma Digitalis.
Digitalis Leaves dried 2 oz.; Linseed Meal 2 oz.; water 1 pint. Boil the leaves with the water for 10 minutes and add the linseed gradually. *London.*

Cataplasma Fermenti.
Beer Yeast 6 fl. oz.; Flour 14 oz.; water at 100° F., 6 fl. oz. *Fever.*

Cataplasma Iodi.
Linseed poultice sprinkled with 2 drs. or more of Tincture of Iodine. *Fever. Women.*

Cataplasma Lini.
Linseed Meal 4 oz.; boiling water 10 oz. *London.*

Cataplasma Micæ Panis.
Grated bread, boiling water, of each sufficient. *London.*
Soft crumb of bread, boiling water, of each *q. s. British Skin.*

Cataplasma Plumbi et Opii.
Linseed poultice, substituting Lead and Opium lotion for half of the water. *Fever. Women.*

Cataplasma Sinapis.
Mustard 2 oz.; Linseed Meal 2 oz.; tepid water 8 oz. *Fever.*

Made with Mustard and water only, or add a sufficiency of Mustard to a Linseed Meal poultice. *London.*

Cataplasma Sodæ Chlorinatæ.
Make a Linseed poultice, substituting Solution of Chlorinated Soda for one-half of the water. *Fever.* (For one-third of the water. *London.*)

Cataplasma Thymolis.
Linseed Meal 2 oz.; Thymol 10 grs.; Olive Oil 2 drms.; Boiling water 5 oz. Dissolve the Thymol in the oil. *British Skin.*

Cataplasma Zingiberis.
Powdered Ginger 1 oz.; Powdered Capsicum ½ oz.; 30 to 60 grs. to be sprinkled on a Linseed Meal poultice. *Middlesex.*

CAUSTICA.

Causticum Acidi Arseniosi.
Arsenious Acid 20 grs.; Vermilion 60 grs.; Benzoated Lard 1 oz. *British Skin.*

Causticum Acidi Carbolici.
Carbolic Acid 1 oz.; water 1 drm. *British Skin.*

Causticum Acidi Chromici.
Chromic Acid 1 oz.; water 1 oz. *British Skin.*

Causticum Argenti Nitratis.
Nitrate of Silver 30 grs.; Rectified Spirit 1 oz. *British Skin.*

Causticum Depilatorium.
Sulphide of Barium 90 grs.; Oxide of Zinc 360 grs.; Carmine 1 gr. Used as a paste with water. *British Skin.*

Causticum Iodi.
Iodine 180 grs.; Iodide of Potassium 60 grs.; Rectified Spirit 1 oz. *British Skin.*

Causticum Pasta Londinensis.

Caustic Soda, Unslaked Lime, equal parts reduced to a fine powder in a warm mortar and bottle it, to be made into a paste with water before use. *Throat.*

Causticum Potassæ c. Calce.

Fused Potassa ½ oz.; Quicklime ½ oz.; made into a paste with Rectified Spirit. *British Skin.*

Causticum Sabinæ.

Savin in powder 380 grs.; Oxide of Zinc 60 grs. *British Skin.*

Caustica Tela Zinci Chloridi.

No. 1. Chloride of Zinc in fine powder. Wheaten Flour, equal parts, rubbed together to form a mass; place it in a drying closet till stiff enough to form cylinders. *Throat.*

No. 2 consists of Chloride of Zinc 2 pts.; Wheaten Flour 3 pts. *Throat.*

No. 3. Chloride of Zinc 1 pt.; Wheaten Flour 2 pts. *Throat.*

Causticum Zinci Chloridi.

Chloride of Zinc, Oxide of Zinc, equal parts. *London Ophthalmic.* (Made into paste with water.) *British Skin.*

Or Chloride of Zinc 480 grs.; Wheat Flour 180 grs.; water 1 oz.; mix and apply the heat of a water bath for a few minutes only until a paste is formed. *London Ophthalmic.*

Causticum Zinci Iodati.

Iodide of Potassium 240 grs.; Iodine 480 grs.; Distilled Water 3 drachms. Dissolve by trituration in a glass mortar. Add the above drop by drop to the following solution: Sulphate of Zinc 200 grs.; Distilled Water 140 minims; dissolve.

Allow the mixture to stand for six hours, then decant the liquid from the sediment, and preserve in a well-stoppered bottle. *Throat.*

CHARTA.
(*See also* FUMI.)

Charta Nitratis Potassæ.
> Nitrate of Potash 1 oz.; Distilled Water 4 oz.;, dissolve; imbue pieces of thin bibulous paper 8 inches square in this solution, and dry. One to be burnt for each inhalation. *Middlesex*.
>
> Nitrate of Potash 4 oz.; boiling water 10 oz.; dissolve; pour into a tray; draw sheets of best white blotting paper through the solution, and hang up to dry. About 8 inches square to be burnt for a dose. *Royal Chest*.

COLLODIA.

Collodium Stypticum.
> Rectified Spirit 3 drms.; Ether 5 drms. Mix and dissolve Tannic Acid 80 grs.; then add Pyroxylin 60 grs. Shake occasionally during 48 hours, and then decant the clear solution. *London*.

COLLUNARIA.
(NASAL DOUCHES.)

Collunarium Acidi Carbolici.
> Carbolic Acid ½ gr.; Glycerine 20 grs.; water to 1 oz. *Throat*.

Collunarium Acidi Carbolici c. Soda et Borace.
> Carbolic Acid 4 grs.; Bicarbonate of Soda 12 grs.; Borax 12 grs.; water 1 oz. *Throat*.

Collunarium Acidi Tannici.
> Tannic Acid 3 grs.; water 1 oz. *London. Throat*.

Collunarium Aluminis.
> Alum 4 grs.; water 1 oz. *London. Throat*.

Collunarium Bituminis Comp.
> Compound Liquor of Bitumen 10 mins.; water to 1 oz. *London*.

Collunarium Potassæ Permanganatis.
Solution of Permanganate of Potash 6 mins.; water to 1 oz. *Throat.*
Solution of Permanganate of Potash 10 mins.; water to 1 oz. *London.*

Collunarium Quiniæ.
Sulphate of Quinine ½ gr.; water 1 oz. *Throat.*

Collunarium Sodæ.
Bicarbonate of Soda 30 grs.; water 1 oz. *Throat.*

Collunarium Zinci Sulphatis.
Sulphate of Zinc ½ gr.; water 1 oz. *Throat.*

Collunarium Zinci Sulpho-carbolatis.
Sulpho-carbolate of Zinc 2 grs.; water 1 oz. *Throat.*

COLLYRIA.

Collyrium Acidi Boracici.
Boracic Acid 5 grs.; Distilled Water 1 oz. *Middlesex.*

Collyrium Aluminis.
Alum 4 grs.; water 1 oz. *London. Middlesex. Samaritan.*
Alum 3 grs.; water to 1 oz. *Guy's. Royal Free.*

Collyrium Aluminis c. Atropia.
Sulphate of Atropia 1 gr.; Alum 24 grs.; Distilled Water 8 oz. *Middlesex.*

Collyrium Aluminis c. Belladonna.
Extract of Belladonna 5 grs.; Alum 3 grs.; water 1 oz. *London.*

Collyrium Argenti Nitratis.
Nitrate of Silver 1 gr.; Distilled Water 1 oz. *City Chest. Middlesex.*

Collyrium Argenti Nitratis Fortius.
Nitrate of Silver 4 grs; Distilled Water 1 oz. *King's.*

Collyrium Argenti Nitratis Mitius.
Nitrate of Silver 2 grs.; Distilled Water 1 oz. *King's.*

Collyrium Atropiæ Sulphatis.
Sulphate of Atropia $\frac{1}{8}$ gr.; water 1 oz. *British Skin.*
Sulphate of Atropia 2 grs.; water 1 oz. *King's.*

Collyrium Belladonnæ.
Belladonna Extract 8 grs.; water 1 oz. *Guy's.*

Collyrium Boracis.
Borax 10 grs.; water 1 oz. *Middlesex.*
Borax 5 grs.; Diluted Hydrocyanic Acid 3 mins.; water to 1 oz. *Royal Free.*

Collyrium Cupri Compositum.
Aluminated Copper 5 grs.; Camphor water 1 drm.; water to 1 oz. *Royal Free.*

Collyrium Cupri Sulphatis.
Sulphate of Copper 2 grs.; water 1 oz. *King's. British Skin. London.*

Collyrium Hydrargyri Perchloridi.
Corrosive Sublimate $\frac{1}{8}$ gr.; water 1 oz. *British Skin.*

Collyrium Plumbi.
Solution of Subacetate of Lead 60 mins.; Elder-flower Water 6 oz. *Samaritan.*

Collyrium Plumbi Acetatis.
Acetate of Lead 2 grs.; Diluted Acetic Acid $\frac{1}{2}$ min; water 1 oz. *Royal Free.*
Acetate of Lead 2 grs.; Diluted Acetic Acid 1 min.; distilled water 1 oz. *Middlesex.*

Collyrium Zinci Acetatis.
Acetate of Zinc 2 grs.; water 1 oz. *Samaritan.*

Collyrium Zinci Chloridi.
Chloride of Zinc 1 gr.; water 1 oz. *London.*
Chloride of Zinc 2 grs; water 1 oz. *Guy's.*

Collyrium Zinci c. Opio.
> Sulphate of Zinc 2 grs.; Tinct. Opium 20 mins.; water 1 oz. *King's.*

Collyrium Zinci Sulphatis.
> Sulphate of Zinc 1 gr.; water 1 oz. *London. Middlesex.*
> Sulphate of Zinc 2 grs.; water 1 oz. *British Skin. City Chest. Guy's. Royal Free. St. George's.*

CONFECTIONES vel ELECTUARIA.

Confectio Cubebæ Co.
> Cubebs 60 grs.; Balsam of Copaiba ½ drm.; Treacle ½ drm. Dose 2 drms. *St. Mary's.*
> Cubebs 30 grs.; Copaiva 20 mins.; Bicarbonate of Soda 10 grs.; Nitrate of Potash 5 grs. *London.*

Confectio Cubebæ et Copaibæ.
> Cubebs 15 grs.; Copaiba 15 mins.; mix. *University.*

Confectio Ferri Tartarati.
> Tartarated Iron 30 grs.; Acid Tartrate of Potash 180 grs.; Ginger 5 grs.; Treacle to 1 fl. oz. Dose 2 teaspoonfuls. *St. Bartholomew's.*

Confectio Guaiaci Co.
> Guaiacum Resin 10 grs.; Carbonate of Magnesia 10 grs.; Sublimed Sulphur 15 grs.; Ginger 5 grs.; Treacle by weight 1 drm. Dose 1 to 2 drms. *London.*

Confectio Jalapæ.
> Powdered Jalap ¼ oz.; Powdered Senna 2 oz.; Ginger 60 grs.; Treacle 8 oz. Dose 60 to 120 grs. *St. George's.*
> Comp. Powder of Jalap 120 grs.; Treacle to 1 fl. oz. Dose 2 to 4 teaspoonfuls. *St. Bartholomew's.*

Confectio Jalapæ et Sulphuris.
> Confection of Jalap 5 oz.; Precipitated Sulphur 1 oz. Dose 60 to 120 grs. *St. George's.*

Confectio Olei Santali Flavi Comp.
> Oil of Yellow Sandal Wood 20 mins.; Cubebs 30 grs.; Bicarbonate of Potash 5 grs.; Nitrate of Potash 5 grs.; Dose 1 to 2 drms. *London.*

Confectio Potassæ Tartratis Acidæ.

Acid Tartrate of Potash 240 grs.; Powdered Ginger 5 grs.; Treacle to 1 fl. oz. Dose 1 to 2 teaspoonfuls. *St. Bartholomew's.*

Acid Tartrate of Potash 20 grs.; Honey 60 grs. *Charing Cross.*

Confectio Rutæ.

Rue 1½, Caraway 1½, Bay Berries 1½, Black Pepper ¼, all in powder; Prepared Sagapenum ½; Honey 16; water *q. s.*; liquefy the last three ingredients by heat, and add the powders. *St. George's.*

Used in ENEMA RUTÆ.

Confectio Sennæ.

Powdered Senna 120 grs.; Jalap 40 grs.; Ginger 10 grs.; Treacle 1 fl. oz. Dose 1 to 2 drms. *St. Mary's.*

Powdered Senna 80 grs.; Jalap 20 grs.; Ginger 20 grs.; Acid Tartrate of Potash 20 grs.; Powdered Liquorice 20 grs.; Coriander 20 grs.; Treacle 1 oz. Dose 1 to 2 drms. *Middlesex.*

(Great Northern is the same form as Middlesex, Coriander being omitted.)

Confectio Sennæ Co.

Confection of Senna ½ drm.; Confection of Sulphur ½ drm. *Samaritan.*

Confectio Sennæ et Jalapæ.

Powdered Senna 20 grs.; Powdered Jalap 10 grs.; Acid Tartrate of Potash 20 grs.; Syrup of Ginger 1 drm. Dose ½ to 1 drm. *London.*

Confectio Sennæ et Sulphuris.

Confection of Senna 6 drms.; Sulphur 1 drm. Dose 1 to 2 drms. *St. Mary's.*

Confection of Senna 75 grs.; Sublimed Sulphur 15 grs.; Acid Tartrate of Potash 15 grs. *University.*

Confection of Senna P.B., and Confection of Sulphur. Hosp. Pharm. Equal parts. Dose 1 to 2 drms. *London.*

Confection of Senna and Confection of Sulphur equal parts. *Guy's.*

Confectio Sulphuris.
Sulphur ½ oz.; Acid Tartrate of Potash ½ oz.; Confection of Senna 2 oz.; Treacle *q. s.* Dose 1 to 2 drms. *Middlesex.*
Sulphur 4 drms.; Acid Tartrate of Potash 1 drm.; Treacle 4 drms. Dose 1 to 2 teaspoonfuls. *London Ophthalmic.*
Sulphur 2 drms.; Acid Tartrate of Potash 1 drm.; Treacle 6 drms. Dose 1 to 2 drms. *London.*
Sulphur 200 grs.; Acid Tartrate of Potash 50 grs.; Treacle to 1 fl. oz. Dose 1 to 2 teaspoonfuls. *St. Bartholomew's.*

Confectio Sulphuris Co.
Sulphur 100 grs.; Acid Tartrate of Potash 25 grs.; Guaiacum Resin 80 grs.; Rhubarb 40 grs.; Nutmeg 40 grs.; Treacle to 1 oz. Dose 2 to 4 teaspoonfuls. *St. Bartholomew's.*
Precipitated Sulphur 120 grs.; Acid Tartrate of Potash 30 grs.; Treacle 6 oz. *British Skin.*

Confectio Sulphuris Fortior.
Sublimed Sulphur 30 grs.; Jalap 4 grs.; Tincture of Senna 8 mins.; Oil of Peppermint ¼ min.; Treacle to 1 drm. *Royal Chest.*

DECOCTA.

Decoctum Aloes.
Extract of Socotrine Aloes 3 oz.; Myrrh 2¼ oz.; Carbonate of Potash 1½ oz.; Water 1½ gallon. Boil for 5 minutes and add Cardamom seeds 2 oz.; Caraway Fruit 2 oz.; Cinnamon 4 oz. All the above in powder. Macerate 3 days, strain, and make up with water to 2 gallons. *Royal Free.*

Decoctum Aloes Co.
Cape Aloes 120 grs.; Myrrh 90 grs.; Carbonate of Potash 60 grs.; Commercial Extract of Liquorice 1 oz.; Tincture of Ginger ¾ oz.; Rectified Spirit 3 oz.; Distilled Water *q. s.*; boil the solids in 20 oz. of the water 5 minutes; when cool, add

the Spirit and Tincture; then strain and make
up with water to 30 oz. *St. Bartholomew's.*
Socotrine Aloes 120 grs.; Myrrh 90 grs.; Carbonate
of Potash 60 grs.; Extract of Liquorice 2 oz.;
Tincture of Ginger 6 drms.; Rectified Spirit 3
oz.; Distilled Water *q. s.*; boil the solids in 20
oz. of the water 5 minutes; when cool, add the
Spirit and Tincture; then strain and make up
with water to 30 oz. Dose ½ to 2 oz. *London.*

Decoctum Cinchonæ.
Pale Cinchona Bark 1¼ oz.; water 20 oz. Boil for
20 minutes and strain. Dose 1 to 2 oz. *London.*

Decoctum Gallæ.
Galls bruised 2½ oz.; water 40 oz.; boil to 20 oz.
St. George's.

Decoctum Glycyrrhizæ Co.
Liquorice Root 1½ oz.; Aniseeds ½ oz.; water to 20
oz.; boil 15 minutes and strain. Dose 1 to 3 fl.
oz. *Consumption.*

Decoctum Hemidesmi Co.
Hemidesmus Root 2½ oz.; Sassafras Chips ¼ oz.;
Guaiacum Turnings ¼ oz.; Commercial Extract
of Liquorice ¼ oz.; Mezereon Bark 60 grs.;
boiling Distilled Water 30 oz.; digest 1 hour
and boil 10 minutes; when cool make up to 20
oz. Dose 1 oz. *St. Bartholomew's.*

Decoctum Hordei Gummosum.
Gum Acacia 1 oz.; Decoction of Barley 40 oz. *St.
George's.*

Decoctum Hordei Tartarizatum.
Acid Tartrate of Potash 60 grs.; Decoction of Barley 30 oz.; boil till dissolved. *St. George's.*

Decoctum Jaborandi.
Jaborandi leaves bruised 200 grs.; Water 20 oz.
Boil 20 minutes in a covered vessel and strain.
Dose 1 oz. *London.*

Decoctum Papaveris. *See* Fotus Papaveris.

Decoctum Quercus.
: Oak Bark bruised 1 oz.; water 20 oz.; boil 15 minutes in a covered vessel and strain. *London Ophthalmic.*

Decoctum Tritici Repentis.
: Creeping Couch Grass 1 oz; Water 20 oz. Boil 10 minutes in a covered vessel and strain. Dose 1 to 2 oz. *London.*

ELECTUARIA. *See* CONFECTIONES.

EMPLASTRA.

Emplastrum Belladonnæ.
: Alcoholic Extract of Belladonna Root 1 drm.; Lead Plaster 19 drms. *University.*

Emplastrum Depilatorium.
: Resin 1 oz.; Tar 110 grs.; Oil of Turpentine ½ drm; Lard 14 grs. *British Skin.*

Emplastrum Galbani c. Cantharide.
: Galbanum Plaster 7 oz.; Cantharides Plaster 1 oz. *St. George's.*

Emplastrum Hydrargyri Compositum.
: Mercury 120 grs.; Oil of Turpentine 20 mins.; Yellow Wax 20 grs.; Resin 40 grs.; Prepared Styrax 60 grs.; Lead Plaster 1 oz. *British Skin.*

Emplastrum Iodi Co.
: Iodine 2 drms.; Iodide of Potassium 3 drms.; Lead Plaster 16 oz.; Opium Plaster 6 oz. *St. George's. Women.*

Emplastrum Rubrum.
: Red Oxide of Lead 45 grs.; Vermilion 25 grs.; Lead Plaster 1 oz. *British Skin.*

ENEMATA.

Enema Assafœtidæ.
Tincture of Assafœtida 4 drms.; Decoction of Barley 20 oz. *Fever.*
Tincture of Assafœtida 4 drms.; Mucilage of Starch 4 oz. *London.*

Enema Calcis c. Ferro.
Tincture of Perchloride of Iron 20 mins.; Lime Water 1 oz. *British Skin.*

Enema Catharticum.
Sulphate of Magnesia 1 oz.; Olive Oil 2 oz.; Decoction of Barley 18 oz. *Fever.*

Enema Chloral.
Hydrate of Chloral 40 grs.; Mucilage of Starch 20 oz. *London.*

Enema Cibi.
Warm Beef-tea 2 fl. oz.; one egg. Mix thoroughly. Brandy or Opium to be added as required. *Guy's.*

Enema Colocynthidis Co.
Colocynth Pulp 60 grs.; water 12 oz.; boil 10 minutes, strain, and add Salt ½ oz. *Guy's.*

Enema Commune.
Chloride of Sodium 1 oz.; Decoction of Barley 20 oz. *Westminster.*
Chloride of Sodium 1 oz.; Decoction of Barley (tepid) 20 oz. *St. Thomas's.*
Common Salt 1 oz.; Oatmeal gruel 12 oz. *Guy's.*

Enema Nutriens.
Jelly Beef tea 3 oz.; Yolk of one egg. Beat up the egg and mix with the beef tea. Port wine Brandy or Opium to be added as ordered. *Middlesex.*
Cooked beef, mutton or chicken, 3 oz. 7 drms.; Sweetbread 1 oz. 7 drms.; Fat 6 drms.; Brandy 2 drms.; Water 3 oz.; These ingredients, mixed together, will measure 9 oz. The meat, sweetbread and fat must be first passed through a fine mincing-machine, and then rubbed up, with

the water gradually added, to make a thick paste. It is well to warm the mass to a temperature of 100° Fahr. shortly before using it. The enema should be given at a temperature of 90° to 95°, and *ought not to be administered more than twice in the twenty-four hours.* The rectum should be washed out twice or thrice a week with tepid water, three or four hours before giving the nutritive injection. Half a drachm of pancreatine (Savory and Moore) or 4 oz. of pancreatic emulsion may be used if the fresh pancreas cannot be obtained. In the latter case the fat, brandy, and water should be omitted. *Throat.*

Warm special beef-tea 2 oz.; one egg. Beat up the egg and mix with the beef-tea. Port wine, Brandy, or Opium to be added as ordered. *London.*

Port wine ½ oz.; Glycerine 1 drm.; Strong Beef Tea (Brand's) to 2 or 3 oz. *Women.*

Enema Nutriens (No. 1).

Strong Beef-tea 5 oz.; Pepsin (Pigs) 10 grs.; Dilute Hydrochloric Acid 20 mins. Digest the Beef Tea, Pepsin, and Acid for 4 hours at a temperature of 100° F., and then neutralize with Carbonate of Soda. *University.*

Enema Nutriens (No. 2).

Strong Beef Tea 3 oz.; One egg. Brandy ½ oz.; Pepsin (Pigs) 12 grs.; Dilute Hydrochloric Acid 24 mins.; Carbonate of Soda a sufficiency. Let the egg be well beaten and mixed with the Beef Tea, Pepsin, and Acid, then digest for 6 hours at a temperature of 100° F., finally neutralize with Carbonate of Soda, and add the Brandy. *University.*

Enema Nutriens cum Peptone.

Lean Beef finely minced 1 lb.; Water 20 oz.; Liquor Pancreaticus 6 drms.; Mix beef and water, simmer gently 1½ hour, strain into a covered jug. Beat the residue into a paste and add it to the jug. When it has cooled to 140° F. add the Pancreatic solution and stir well together. Put it in a warm place for 2 hours, then boil for 3 minutes and strain Not more than 4 oz. to be injected at a time. *London.*

21

Enema Olei Olivæ. *See* Enema Oleosum.

Enema Olei Ricini.
Castor Oil 2 oz.; Decoction of Barley (tepid) 8 oz. *St. Thomas's.*
Castor Oil 2 oz.; Yolk of 1 Egg; Decoction of Barley (tepid) to 10 oz.; mix. *Charing Cross.*
Castor Oil 2 oz.; Soft Soap 1 oz.; warm water 20 oz. *Westminster.*
Castor Oil 1 oz.; warm Mucilage of Starch 11 oz. *Guy's.*
Castor Oil 2 oz.; Mucilage of Starch 18 oz. *St. Bartholomew's.*
Castor Oil 1 oz.; Simple enema to 20 oz. *Middlesex.*
Castor Oil 1 oz.; Mucilage of Starch 8 oz. *London.*

Enema Olei Ricini c. Assafœtidâ.
Castor Oil 4 drms.; Tincture of Assafœtida 4 drms.; Mucilage of Starch to 12 oz. *Fever.*

Enema Oleosum.
Olive Oil 4 oz.; Decoct. of Oats 12 oz. *St. George's.*
Olive Oil 4 oz.; Simple Enema to 20 oz. *Middlesex.*
Olive Oil 4 oz.; Mucilage of Starch 16 oz. *London.*
Olive Oil 4 oz.; Warm Mucilage of Starch 8 oz. *Guy's.*

Enema Opii.
Tincture of Opium 15 mins.; Mucilage of Starch 2 oz. *Fever.*

Enema Plumbi c. Opio.
Tincture of Opium 20 mins.; Acetate of Lead 9 grs.; Dilute Acetic Acid 15 mins.; Distilled Water 3 oz. *Fever.*

Enema Rutæ.
Confection of Rue 180 grs.; Infusion of Chamomile 16 oz. *St. George's.*
Oil of Rue 20 mins.; Mucilage of Acacia 2 drms.; Simple Enema to 6 oz. *Middlesex.*
Oil of Rue 20 mins.; Mucilage 2 drms.; Mucilage of Starch 6 oz. *Women.*
Oil of Rue 20 mins.; Mucilage of Starch 6 oz. *Westminster.*

Enema Sanguinis Exsiccati.

No. 1. Defirbrinated Bullock's Blood desiccated 180 grs.; Milk warmed to 100° F., 3 oz. Shake until dissolved. *London.*

No. 2. Defirbrinated Bullock's Blood desiccated 2 oz.; Milk warmed to 100° F., 8 oz. Shake until dissolved. *London.*

Brandy, Port Wine or Opium to be added to either as ordered.

Enema Saponis.

Soft Soap ¾ oz.; boiling water 16 oz. *St. George's.*
Soft Soap ¾ oz.; warm water to 20 oz. *St. Bartholomew's.*
Soft Soap 1 oz.; tepid water 30 oz. *Charing Cross.*
Soft Soap 1 oz.; boiling water 16 oz.; to be used warm. *Guy's.*
Soft Soap q. s.; tepid water 40 oz. *Fever.*
Soft Soap 1 oz.; warm water 20 oz. *Westminster.*
Soft Soap 1 oz.; warm water 19 oz. *London.*

Enema Sennæ Co.

Misturæ Nigræ 4 oz.; Decoction of Barley (tepid) 8 oz. *St. Thomas's.*

Enema Simplex.

Linseed Oil 2 oz.; Decoction of Barley 10 oz. *London.*
Soft Soap 1 oz.; boiling water 20 oz. *Middlesex.*

Enema Sodii Chloridi.

Chloride of Sodium 1 oz.; warm water 16 oz. *St. George's.* (*London*, 20 oz.)
Chloride of Sodium 1 oz.; Decoction of Barley 20 oz. *Westminster.*

Enema Spiritus Vini Gallici.

Brandy 1 oz.; Strong Beef Tea 3 oz. *Fever.*

Enema Terebinthinæ.

Oil of Turpentine 1 oz.; Mucilage of Starch 20 oz. *Fever.*
Oil of Turpentine ½ oz.; Castor Oil 1 oz.; Yolk of 1 egg; Infusion of Linseed to 20 oz.; mix. *Charing Cross.*

Oil of Turpentine ½ oz.; Castor Oil 1 oz.; Decoction of Barley or Mucilage of Starch 20 oz. *Middlesex*.

Enema Vulgare. *See* Enema Commune.

ESSENTIA.
Essentia Cassiæ.
: Oil of Cassia 1 oz.; Rectified Spirit 4 oz. Dose 10 to 20 mins. *London.*

EXTRACTA.
Extractum Ergotæ Ammoniatum Liquidum.
: Ergot powdered 1 lb., Aromatic Spirit of Ammonia 10 oz.; Glycerine 2 oz.; Water *q. s.* Macerate the first three together in a percolator for 12 hours, then continue the percolation with water until 13 oz. have been collected. Digest the marc in 40 oz. water at 160° F. for 2 hours, then press, strain, and evaporate the liquor to 3½ oz. Mix it with the percolate, set aside 12 hours, and filter. Dose 10 to 30 mins. *London.*

Extractum Malti c. Oleo Morrhuæ.
: Cod Liver Oil 1 oz.; Gum Acacia in powder 120 grs.; Cassia water 1 oz. Make an emulsion and add Extract of Malt 2 fl. oz. Mix. Dose 2 to 4 drms. *London.*

Extractum Quebracho Liquidum.
: Quebracho bark in powder 16 oz.; Glycerine 4 oz.; Water 4 oz.; Rectified Spirit 8 oz.; Proof Spirit a sufficiency, boiling Distilled water ½ gal. Add to the Quebracho the Glycerine, Water, and Rectified Spirit mixed together, and set aside for 4 days, then allow it to drain, and continue the percolation with Proof Spirit until 12 fl. oz. have been collected; then add to the marc the boiling Distilled water, and macerate for 24 hours, press, filter, and evaporate the liquor to 4 fl. oz.; mix this with the percolate, and filter. Dose 20 to 60 mins. *London.*

FOMENTUM. *See* Fotus.

FOTUS.

Fotus Belladonnæ.

Extract of Belladonna 60 grs.; hot Distilled Water 20 oz. *Great Northern. London Ophthalmic. St. Bartholomew's. Westminster. Westminster Ophthalmic.*

Extract of Belladonna 60 grs.; water 20 oz. *Middlesex.*

Extract of Belladonna 3 grs.; water to 1 oz.; to be used warm. *London.*

Extract of Belladonna 40 grs.; water to 1 oz.; ½ oz. in a teacupful of boiling water. *Royal Free.*

Fotus Papaveris.

Bruised Poppy Capsules 2 oz.; water 30 oz.; boil ten minutes and strain, and make up to 20 oz. *St. Bartholomew's.*

Capsules 1 oz.; boiling water 20 oz.; boil fifteen minutes and strain. *London Ophthalmic. Middlesex. Westminster. Westminster Ophthalmic.*

Capsules sliced, 3 or more.; water 2 pints.; boil for a quarter of an hour and strain. *Fever. Women.*

Poppy capsules freed from seeds and bruised 1 oz.; Water 40 oz.; boil for 15 minutes and strain. *London.*

Fotus Terebinthinæ.

Hot damp flannel sprinkled with 2 fl. drms. Oil of Turpentine, covered with gutta percha tissue. *Fever. London. Women.*

FUMI.
(FUMING INHALATIONS.)

Fumus Potassæ Nitratis (Nitrated Papers).

No. 1.
Nitrate of Potash 30 grs.; water 1 oz.; dissolve.

No. 2.
Nitrate of Potash 45 grs.; water 1 oz.; dissolve.

No. 3.
Nitrate of Potash 60 grs.; water 1 oz.; dissolve. *Throat.*

Saturate white blotting-paper in either of these solutions, and dry it.
Cut the paper into pieces 3 inches long, ½ inch broad, which enables the medical attendant to order a definite quantity. Light a paper, drop it into the fuming inhaler, or any cylindrical vessel, and inhale the smoke. From 1 to 6 papers may be used in succession for each inhalation.
Antispasmodic.

Note.—A particular character may be given to these papers by the addition of various volatile principles. Thus Camphor and Cassia increase their powers, whilst Benzoin, Santal, and Sumbul reduce their action and make them less irritating. The medium strength (No. 2) is generally employed in these cases, and the method of preparing them is to moisten the papers with the Tincture, or in the case of essential oils with a solution of the oil (1 drachm in 9 drachms of Rectified Spirit), and then to expose the papers for a few minutes to allow the spirit to pass off.
These papers should be kept in tinfoil.

NITRATED PAPERS.

With Comp. Tinct. Benzoin.
,, Oil Cassia.
,, ,, Cinnamon.
,, ,, Santal.
,, Sp. Camphor.
,, Tinct. of Sumbul.

Pulvis Anti-Asthmaticus. *See* p. 206.

FUMIGATIO.
(See also VAPORES.)

Fumigatio e Chartâ Potassæ Nitratis.
(Inhalation of smoke from ignited Nitrated Papers.)
Nitrate of Potash ¼ oz.; Water 1 fl. oz.; mix. Saturate white blotting paper in the solution, and dry it. Cut the paper into pieces 3 inches long and 1 inch broad. Light a paper, drop it into a suitable vessel, and let the smoke be inhaled. From one to six pieces may be used in succession at each inhalation. *London.*

Fumigatio Chlori.
Chloride of Sodium 1 oz.; Black Oxide of Manganese ¾ oz. Chlorine is evolved on the addition of Sulphuric Acid. *Middlesex.*

GARGARISMATA.

Gargarisma.
Oxymel 1 oz.; Decoction of Barley 5 oz. *St. George's.*

Gargarisma Aceticum.
Acetic Acid 15 mins.; Oxymel 30 mins.; Distilled Water to 1 fl. oz. *St. Bartholomew's. City Chest.*

Gargarisma Acidi Acetici.
Acetic Acid 15 mins.; Glycerine 18 mins.; water to 1 oz. *Throat.*
Stimulant and Antiseptic.

Gargarisma Acidi Carbolici.
Saturated cold aqueous solution of Carbolic Acid (= 48 grs. Crystals) 1½ oz.; water 8½ oz. *Fever.*
Carbolic Acid (No. 2 Calvert) 2 grs; Hot water 2 drms.; dissolve and add water to 1 oz. *London.*
Carbolic Acid (No. 2 Calvert) 3 grs.; Water 1 oz. *Guy's.*
Carbolic Acid 20 grs.; Glycerine ½ oz.; water to 10 oz. *Throat.*
Glycerine of Carbolic Acid 2½ drms.; Glycerine 3 drms.; water to 20 oz. *London Ophthalmic.*
Stimulant and antiseptic.

Gargarisma Acidi Hydrochlorici.
Dil. Hydrochloric Acid 60 mins.; Decoction of Yellow Bark to 4 oz. *City Chest. London. Royal Chest. Westminster.*
Dil. Hydrochloric Acid 60 mins.: Decoction of Pale Bark 6 oz. *St. Bartholomew's.*
Dil. Hydrochloric Acid 60 mins.; water to 3 oz. *Royal Free.*
Dil. Hydrochloric Acid 60 mins.; water to 6 oz. *Guy's.*
Dil. Hydrochloric Acid 1 drm.; Honey ¼ oz.; water to 3 oz. *Samaritan.*

Dil. Hydrochloric Acid 1 drm.; Glycerine 2 drms.; water 5 oz. *Throat.*
Hydrochloric Acid 1 drm.; water 15 oz. *Women.*
Hydrochloric Acid 1 drm.; water to 20 oz. *St. Thomas's.*
Stimulant.

Gargarisma Acidi Nitrici.
Dil. Nitric Acid 20 mins.; water to 1 oz. *Royal Free.*
Nitric Acid 1 fl. drm.; Syrup 1 oz.; water to 12 oz. *St. Mary's.*
Nitric Acid 1 fl. drm.; water to 20 oz. *St. Thomas's.*

Gargarisma Acidi Sulphurici.
Dil. Sulphuric Acid 8 mins.; Treacle 20 grs.; water to 1 oz. *Royal Free.*

Gargarisma Acidi Tannici.
Tannic Acid 50 grs.; Glycerine 2 drms.; water to 5 oz. *Consumption.*
Tannic Acid 10 grs.; Water 1 oz. *Guy's.*
Tannic Acid 1 drm.; Rectified Spirit ½ drm.; Camphor water to 5 oz. *Throat.*
Tannic Acid 60 grs.; Rectified Spirit 30 mins.; water to 6 oz. *London.*
Tannic Acid 1 drm.; Glycerine 3½ drms.; Compound Infusion of Roses to 3 oz. *Fever.*
Glycerine of Tannic Acid 1 drm.; water to 1 oz. *Royal Chest. Westminster.*
Glycerine of Tannic Acid 30 mins.; water to 1 oz. *City Chest. University.*
Glycerine of Tannic Acid ½ drm.; Rose water to 1 oz. *Great Northern.*
Glycerine of Tannic Acid 24 mins.; water to 1 oz. *Middlesex.*
Glycerine of Tannic Acid 2 drms.; water to 1 oz. *British Skin.*
Glycerine of Tannic Acid 3 drms.; Dil. Hydrochloric Acid 2 drms.; water 20 oz. *London Ophthalmic.*
Astringent.

Gargarisma Acidi Tannici et Gallici.
Tannic Acid 360 grs.; Gallic Acid 120 grs.; water 1 oz.; mix. *Throat.*
This preparation is most useful for arresting

hæmorrhage from the uvula or tonsils after excision; the patient should be directed to slowly sip the mixture, or hold it passively in the mouth till the hæmorrhage is stopped.

Gargarisma Æruginis.

Liniment of Verdigris P. L. 1 oz.; water to 4 oz. *St. Thomas's.*
Liniment of Verdigris 1 oz.; Mucilage of Acacia 2 oz.; water to 12 oz. *Middlesex.*
Verdigris 15 grs.; Vinegar 1½ drm.; Honey 3 drms.; Lime Water 20 oz. *St. George's.*

Gargarisma Aluminii Chloridi.

Solution of Chloride of Aluminium, sp. g. 1250, 2 drms.; water 10 oz. *Throat.*
Astringent and Antiseptic.

Gargarisma Aluminis.

Alum 1 drm.; water to 4 oz. *St. George's.*
Alum 1½ drm.; Glycerine ½ oz.; Distilled water 10 oz. *London Ophthalmic.*
Alum 1 drm.; water 6 oz. *Great Northern. London. Royal Chest. St. Mary's.*
Alum 1 drm.; Acid Infusion of Roses 6 oz. *City Chest.*
Alum 1 drm.; Glycerine ½ oz.; water 12 oz. *Women.*
Alum 160 grs.; water to 10 oz. *St. Bartholomew's.*
Alum 50 grs.; Oxymel 2 drms.; water to 5 oz. *Consumption.*
Alum 1 drm.; Honey 4 drms.; water to 4 oz. *University.*
Alum 1 drm.; Tincture of Myrrh 1 drm.; Treacle 2 drms.; water to 6 oz. *Royal Free.*
Alum ½ oz.; Glycerine ½ oz; water to 20 oz. *Charing Cross.*
Alum. 1 drm.; Glycerine 2 drms.; water to 6 oz. *Middlesex.*
Alum 1 drm.; water 7½ oz. *Throat.*
Alum 100 grs.; Diluted Sulphuric Acid 2 drms.; Treacle 6 drms.; water to 10 oz. *Fever.*
Alum 1 drm.; water 8 oz. *St. Thomas's.*
Alum 24 grs.; Treacle 4 drms.; water to 2 oz. *Westminster.*
Alum 1 drm.; water 12 oz. *Guy's.*

Alum 1 drm.; Diluted Sulphuric Acid 3 drms.; water to 20 oz. *King's.*
Mildly astringent.

Gargarisma Aluminis Co.

Alum 160 grs.; Tincture of Myrrh 150 mins.; Distilled Water to 10 fl. oz. *St. Bartholomew's.*
Alum 60 grs.; Syrup of Red Roses 1½ oz.; Decoction of Yellow Cinchona Bark to 12 oz. *Guy's.*
Alum 60 grs.; Dil. Sulphuric Acid 2 drms.; Tincture of Myrrh 4 drms.; water to 12 oz. *London.*

Gargarisma Aluminis c. Acido Tannico.

Alum 60 grs.; Tannic Acid 80 grs.; water 10 oz. *Throat.*
Alum 100 grs.; Tannic Acid 100 grs.; water to 10 oz. *London.*
Astringent.

Gargarisma Astringens.

Tannic Acid 60 grs.; Glycerine 1 oz.; Alum 120 grs.; water 20 oz. *Samaritan.*

Gargarisma Boracis.

Borax 1 oz.; Oxymel 2 oz.; boiling water 14 oz. *St. George's.*
Borax ½ oz.; Oxymel 1 oz.: water to 12 oz. *Middlesex.*
Borax 50 grs.; Honey 2 drms.; water to 5 oz. *Consumption.*
Borax 2 drms.; Honey 60 grs.; water 10 oz. *Samaritan.*
Borax 1 drm.; Treacle 2 drms.; water to 6 oz. *Royal Free.*
Borax 1 drm.; Syrup of Roses 1½ oz.; Barley water to 12 oz. *Guy's.*
Borax 1 drm.; Treacle 10 drms.; water to 5 oz. *Westminster.*
Borax 1 drm.; Treacle 4 fl. drms.; water to 6 oz. *St. Mary's.*
Borax 2 drms.. Tinct. of Myrrh ½ oz.; water to 5 oz. *London.*
Borax 50 grs.; Treacle ½ oz.; water to 5 oz. *Fever.*
Borax 1 drm.; water 4 oz. *Royal Chest. City Chest.*
Borax 1 drm.; Glycerine 1 drm.; Tinct. of Myrrh 1 drm.; water to 2½ oz. *Throat.*
Mildly alkaline astringent.

Gargarisma Boracis c. Myrrhâ.

Glycerine of Borax 5 drms. ; Tinct. of Myrrh 100 mins. ; water to 5 oz. *Consumption.*

Gargarisma Capsici.

Tincture of Capsicum 1 drm.; water 8 oz. *St. George's.*
Tincture of Capsicum 100 mins. ; Dil. Sulphuric Acid 60 mins.; Decoction of Pale Bark to 10 oz. *St. Bartholemew's.*
Tincture of Capsicum 1½ drm.; water to 10 oz. *King's.*
Tincture of Capsicum 1 drm.; water 10 oz. *St. Thomas's.*
Tincture of Capsicum 1 drm.; Dil. Acetic Acid 2 drms. ; water to 6 oz. *Royal Chest.*

Gargarisma Capsici Acidum.

Tincture of Capsicum 3 drms.; Diluted Nitric Acid 3 drms.; Infusion of Red Gum to 12 oz. *London.*

Gargarisma Chlori. (*See also* Vapor Chlori.)

Solution of Chlorine ½ oz. ; water to 6 oz. *Middlesex.*
Solution of Chlorine 1 drm.; water to 6 oz. *London.*

Gargarisma Ferri Aluminis.

Iron Alum 8 grs.; water 1 oz. *Throat.*

Gargarisma Ferri Perchloridi.

Solution of Perchloride of Iron ½ oz.; water to 12 oz. *St. Mary's.*
Solution of Perchloride of Iron 100 mins.; Glycerine 2 drms.; water to 5 oz. *Consumption.*

Gargarisma Hydrargyri.

Corros. Sublimate 5 grs.; Glycerine 200 mins.; Decoction of Cinchona to 10 oz. *Women.*

Gargarisma Hydrargyri Perchloridi.

Corros. Sublim. 2½ grs.; Glycerine ½ oz. ; water to 10 oz. *Throat.*

Corros. Sublim. 4 grs.; water 8 oz. *British Skin.*
Corros. Sublim. 3 grs.; Glycerine 1 oz.; Hydrochloric Acid 12 mins.; water to 12 oz. *St. Mary's.*
Corros. Sublim. 2 grs.; Dil. Hydrochloric Acid 2 drms.; Glycerine 1 oz.; water to 10 oz. *Middlesex.*
Corros. Sublim. 2 grs.; Tinct of Cochineal 2 drms.; water to 20 oz. *London Ophthalmic.*
Corros. Sublim. ½ gr.; Dil. Hydrochloric Acid 5 mins.; Glycerine 1 drm.; water to 1 oz. *Gt. Northern.*
Corros. Sublim. 1 gr.; Dil. Hydrochloric Acid 40 mins.; water to 8 oz. *London.*

Gargarisma Hydrochloricum.

See Gargarisma Acidi Hydrochlorici.

Gargarisma Iodi.

Tincture of Iodine 2 drms.; water to 5 oz. *St. Thomas's.*
Tincture of Iodine ½ drm.; Tinct. of Bark 1½ drm.; water 4 oz. *St. Mary's.*

Gargarisma Krameriæ.

Rhatany Root bruised ½ oz.; water, 100° F., 10 oz. Infuse one hour. *Throat.*
Mildly astringent.

Gargarisma Myrrhæ.

Tincture of Myrrh 1 drm.; water to 3 oz. *Middlesex. Gt. Northern.*

Gargarisma Potassii Bromidi.

Bromide of Potassium 100 grs.; water 10 oz. *Throat.*

Gargarisma Potassæ Chloratis.

Chlorate of Potash 1 drm.; water 3 oz. *British Skin. University.*
Chlorate of Potash ½ oz.; water 12 oz. *Middlesex.*
Chlorate of Potash 1 drm.; Dil. Hydrochloric Acid 40 mins.; water to 4 oz. *Royal Chest.*
Chlorate of Potash 1 drm.; water 5 oz. *Throat.*

Chlorate of Potash 1 drm.; Treacle 10 drms.; water to 5 oz. *Westminster.*
Chlorate of Potash ½ oz.; Dil. Hydrochloric Acid 4 drms.; water to 20 oz. *Charing Cross.*
Chlorate of Potash 1 drm.; water 6 oz. *City Chest. Guy's. London. St. Mary's. Women.*
Chlorate of Potash 1 drm.; Glycerine 1½ drm.; water to 6 oz. *Royal Free.*
Chlorate of Potash 90 grs.; Hydrochloric Acid 40 mins.; Glycerine 1 oz.; water to 20 oz. *London Ophthalmic.*

Gargarisma Potassæ Chloratis Acidum.

Chlorate of Potash 50 grs.; Dil. Hydrochloric Acid 50 mins.; Syrup 2 drms.; water to 5 oz. *Consumption.*
Antiseptic.

Gargarisma Potassæ Permanganatis.

Solution of Permanganate of Potash 6 mins.; water 1 oz. *Throat.*
Solution of Permanganate of Potash 10 mins.; water to 1 oz. *London. St. Mary's.*
Solution of Permanganate of Potash 12 mins.; water to 1 oz. *Fever. University.*

Gargarisma Sodæ Chloratæ.

Chlorinated Solution of Soda 4 drms.; water to 8 oz. *British Skin.*
Chlorinated Solution of Soda 4 drms.; water to 10 oz. *Charing Cross. London Ophthalmic. Throat.*
Chlorinated Solution of Soda 4 drms.; Treacle ¼ oz.; water to 5 oz. *Gt. Northern.*
Chlorinated Solution of Soda 4 drms.; water to 2 oz. *St. Thomas's.*
Chlorinated Solution of Soda 4 drms.; water to 4 oz. *University.*
Chlorinated Solution of Soda 4 drms.; water to 5 oz. *Fever.*
Chlorinated Solution of Soda 4 drms.; water to 12 oz. *London. Middlesex. Royal Free. St. George's.*
Chlorinated Solution of Soda 5 drms.; water to 10 oz. *St. Bartholomew's.*
Chlorinated Solution of Soda 4 drms.; water to 20 oz. *King's.*

Gargarisma Sodæ Sulphocarbolatis.
Sulphocarbolate of Soda 4 grs.; Borax 18 grs.; Glycerine 24 mins.; water to 1 oz. *Throat.*

GLYCERINA.

Glycerinum Acidi Carbolici.
Carbolic Acid 1 oz.; Glycerine 1 oz. *Women.*

Glycerinum Acidi Tartarici.
Synonym.—Vidal's Glycerole.
Tartaric Acid, in fine powder, 22 grs.; Glycerine of Starch 1 oz. Mix thoroughly. Used as a topical stimulant in cases of Lichen circumscriptus and Lichen planus. *British Skin.*

Glycerinum Belladonnæ.
Extract of Belladonna 8 drms.; water 2 drms; Glycerine to 2 oz. *University.*
Extract of Belladonna 1 oz.; water ½ oz.; Glycerine 3½ oz. *London Ophthalmic.*
Extract of Belladonna ½ oz.; water 1 drm.; Glycerine to 1 oz. *London.*
Extract of Belladonna ½ oz.; water 1 drm.; Glycerine 1 oz. *Middlesex.*

Glycerinum Ferri Perchloridi.
Solution of Perchloride of Iron 1 oz.; Glycerine 1 oz. *Middlesex.*

Glycerinum Guaiaci.
Tincture of Guaiacum 30 mins.; Glycerine to 2 drms. *British Skin.*

Glycerinum Pepsinæ.
Pepsine pure 64 grs.; Hydrochloric Acid pure 1 drm.; water 1 oz. Dissolve and add Glycerine pure to 4 oz. (Each drachm contains 2 grains of Pepsine.) Dose 1 to 2 drms. *London.*

Glycerinum Plumbi Subacetatis (Glycerole of Lead).
Solution of Subacetate of Lead 20 oz.; Glycerine 20 oz.; mix and evaporate to 20 oz. *London.*

Acetate of Lead 5 oz.; Oxide of Lead in powder 3½ oz.; Glycerine 20 oz.; Distilled Water 12 oz.; mix and boil together 15 minutes, filter, and evaporate to 20 oz. *London Ophthalmic.*

Glycerinum Plumbi Acetatis.

Acetate of Lead 1 drm.; Glycerine 1 fl. oz. *Guy's.*

Glycerinum Ichthyocollæ.

Isinglass 10 grs.; Glycerine 1 fl. oz. Dissolve with a gentle heat.
Used in cases of Eczema, Lichen, and Ichthyosis. *British Skin.*

Glycerinum Olei Cadi.

Juniper Tar 80 mins.; Glycerine of Starch 1 oz. Mix thoroughly.
Used as a mild topical stimulant in cases of Eczema in the scaly stage. *British Skin.*

Glycerinum Tragacanthæ.

Tragacanth in powder 1 drm.; Glycerine 1 oz.; mix. *University.* (Mix with the aid of heat. *London.*)
Tragacanth 60 grs.; Glycerine ½ fl. oz.; water 1½ drm. Mix and heat for ten minutes in a water-bath. *Throat.*

GOSSYPIA MEDICATA.

Gossypium Acidi Boracici.

Boracic Acid 60 grs.; Glycerine 20 mins.; Water 6 drms.; Cotton Wool, in a thin sheet, 60 grs.; mix the Boracic Acid, Glycerine, and water, and dissolve with the aid of heat. Saturate the wool evenly with the solution, and dry by exposure to the air with a moderate heat. *Throat.*
Antiseptic and disinfectant.

Gossypium Acidi Tannici.

Tannic Acid 30 grs.; Glycerine 10 mins.; Water 6 drms.; Cotton Wool, in a thin sheet, 60 grs. Mix the Tannic Acid, Glycerine, and water.

Dissolve, and saturate the wool evenly with the solution, then dry by exposure to the air with a moderate heat. *Throat.*
Astringent.

Gossypium Aluminis.

Alum 30 grs.; Glycerine 10 mins.; Water 1 oz.; Cotton Wool, in a thin sheet, 60 grs. Dissolve the Alum in the water, add the Glycerine, and saturate the wool evenly with the solution, then dry by exposure to the air with a moderate heat. *Throat.*
Astringent.

Gossypium Camphoræ.

Camphor 30 grs.; Pure Ether 1 oz.; Cotton Wool, in a thin sheet, 60 grs. Dissolve the Camphor in the Ether, and saturate the wool evenly with the solution. Dry by exposure to the air, remove from any artificial light, and keep in a stoppered bottle. This wool should be prepared in a room in which there is neither artificial light nor fire. *Throat.*
Stimulant and antiseptic.

Gossypium Cubebæ.

Tincture of Cubebs 1 oz.; Glycerine 10 mins.; Cotton Wool, in a thin sheet, 60 grs. Mix the Tincture and Glycerine, and saturate the wool evenly with the mixture. Dry by exposure to the air. *Throat.*
In catarrh with excessive secretion.

Gossypium Ferratum.

Moisten cotton wool with glycerine then express strongly, steep the damp wool in a solution of Sulphate of Iron, one part to two parts of water, squeeze out as much as possible of the liquid, and without drying pack the prepared wool into a bottle furnished with a glass stopper. *London.*

Gossypium Ferri Perchloridi.

Solution of Perchloride of Iron $\frac{1}{2}$ oz.; Glycerine 10 mins.; Cotton Wool, in a thin sheet, 60 grs.

Mix the solution and Glycerine and saturate the wool evenly with the mixture. Dry by exposure to the air. *Throat.*
Astringent and styptic.

Note.—In the case of Tannic Acid and Perchloride of Iron Wools, when it is desired to prepare the wools quickly, spirit should be used as a solvent for the Tannic Acid instead of water, and Tincture of Perchloride of Iron in place of the solution.

Gossypium Hamamelis.

Tincture of Hamamelis ½ oz.; Glycerine 10 mins.; Cotton Wool, in a thin sheet, 60 grs. Mix the Tincture and Glycerine, and saturate the wool evenly with the mixture. Dry by exposure to the air. *Throat.*
Astringent.

Gossypium Iodatum.

Cotton wool 10 drms.; Iodine, in fine powder, 1 drm. Into a glass flask having a large mouth furnished with a glass stopper introduce the cotton in small pieces at a time, adding to each about its proportion of Iodine; apply the heat of a sand-bath, and when the expanded air has escaped close the flask and continue the heat until the Iodine is vaporized and the cotton has assumed a dark brown colour. *London. Women.*

Gossypium Iodi.

Tincture of Iodine ½ oz.; Glycerine 10 mins.; Cotton Wool, in a thin sheet, 60 grs. Mix the Tincture and Glycerine, and saturate the wool evenly with the mixture; dry by exposure to the air, and keep in a stoppered bottle. *Throat.*
Stimulant and disinfectant.

Gossypium Iodoformi.

Iodoform 70 grs.; Pure Ether 10 fl. drms. Absolute Alcohol 2 fl. drms.; Glycerine 10 mins.; Cotton Wool, in a thin sheet, 60 grs.; dissolve the Iodoform in the Ether, to which add the Alcohol and Glycerine previously mixed. Saturate the wool evenly with the solution, dry carefully by exposure to the air. This

wool should be prepared in a room in which there is neither artificial light nor fire. *Throat.*
Stimulant and antiseptic.

Gossypium Krameriæ.
Tincture of Rhatany ½ oz.; Glycerine 10 mins.; Cotton Wool, in a thin sheet, 60 grs. Mix the Tincture and Glycerine and saturate the wool evenly with the mixture; dry by exposure to the air. *Throat.*
Astringent.

Gossypium Opii.
Tincture of Opium ½ oz.; Glycerine 10 mins.; Cotton Wool, in a thin sheet, 60 grs. Mix the Tincture and Glycerine, and saturate the wool in the mixture. Dry by exposure to the air. *Throat.*
Sedative.

Gossypium Zinc Chloridi Saturatum.
(No form given.) *Women.*

GUTTÆ.
(For the Eyes and Ears.)

Guttæ Argenti Nitratis.
Nitrate of Silver 2 grs.; Distilled Water 1 oz.
St. Mary's. Royal Free.
Nitrate of Silver 1 gr.; Distilled Water to 1 oz.
Guy's. Gt. Northern. London. London Ophthalmic. Westminster. Westminster Ophthalmic.

Guttæ Argenti Nitratis Fortiores.
Nitrate of Silver 2 grs.; Distilled Water to 1 oz.
London Ophthalmic.
Nitrate of Silver 5 grs.; Distilled Water 1 oz.
London.
Nitrate of Silver 6 grs.; Distilled Water 1 oz.
For the Ear. *Throat.*

Guttæ Argenti Nitratis Fortissimæ.
Nitrate of Silver 12 grs.; Distilled Water 1 oz.
For the Ear. *Throat.*

Guttæ Argenti Nitratis Mitiores.
Nitrate of Silver 3 grs.; Distilled Water 1 oz.
For the Ear. *Throat.*

Guttæ Atropiæ. (*See also below.*)
Sulphate of Atropia ½ gr.; water 1 oz. *Guy's.*
Sulphate of Atropia 1 gr.; Distilled Water 1 oz.
Middlesex. Westminster.
Sulphate of Atropia 2 grs.; water 1 oz. *Gt. Northern.*

Guttæ Atropiæ Fortiores.
Sulphate of Atropia 2 grs.; Distilled Water 1 oz.
Middlesex.

Guttæ Atropiæ c. Alumine.
Sulphate of Atropia 1 gr.; Alum 3 grs.; water 1 oz.
London. London Ophthalmic.

Guttæ Atropiæ Sulphatis.
Sulphate of Atropia 1 gr.; water to 1 oz. *London.
London Ophthalmic. St. Bartholomew's.*
Sulphate of Atropia 2 grs.; water to 1 oz. *Royal Free.*
Sulphate of Atropia 3 grs.; Glycerine 10 mins.; water 1 oz. *St. Mary's.*
No. 1. ⅛ gr. Sulphate to the oz. No. 2. 1 gr. Sulphate to the oz. No. 3. 2 grs. Sulphate to the oz. No. 4. 4 grs. Sulphate to the oz. of water. *Westminster Ophthalmic.*

Guttæ Atropiæ Sulphatis Fortiores.
Sulphate of Atropia 2 grs.; Distilled Water 1 oz.
London Ophthalmic.
Sulphate of Atropia 4 grs.; Distilled Water 1 oz.
London.

Guttæ Atropiæ et Zinci Sulphatis.
Sulphate of Atropia 1 gr.; Sulphate of Zinc 1 gr.; Distilled Water 1 oz. *London. London Ophthalmic. Westminster.*
Sulphate of Atropia ½ gr.; Sulphate of Zinc 1 gr.; Distilled Water 1 oz. *Westminster Ophthalmic.*

Guttæ Cupri Sulphatis.
Sulphate of Copper 2 grs.; water 1 oz. *Guy's. London Ophthalmic. Westminster. Westminster Ophthalmic.*
Sulphate of Copper 1 gr.; water 1 oz. *Middlesex.*

Guttæ Daturinæ.*
Sulphate of Daturine 2 grs.; water 1 oz. *London. London Ophthalmic.*

Guttæ Eserinæ.†
Sulphate of Eserine 2 grs.; water 1 oz. *Gt. Northern. London. London Ophthalmic. Middlesex. St. Bartholomew's.*
Sulphate of Eserine 4 grs.; water 1 oz. *Royal Free.*

Guttæ Homatropinæ.
Hydrobromate of Homatropine 2 grs.; Distilled Water 1 oz. *London.*

Guttæ Hydrargyri Perchloridi.
Corrosive Sublimate ⅛ gr.; water 1 oz. *Westminster. Ophthalmic. Royal Free.*

Guttæ Opii.
Wine of Opium 2 drms.; water to 1 oz. *London Ophthalmic. Westminster. Westminster Ophthalmic.*
Wine of Opium 2 drms.; water 3 drms. *Guy's.*

Guttæ Pilocarpiæ.‡
Nitrate of Pilocarpine 2 grs.; water 1 oz. *London. London Ophthalmic.*

Guttæ Potassii Iodidi.
Iodide of Potassium 4 grs.; water 1 oz. *London Ophthalmic.*
Iodide of Potassium 3 grs.; water 1 oz. *Westminster Ophthalmic.*

---

* From Stramonium.
† From Calabar Bean.
‡ From Jaborandi.

Guttæ Terebinthinæ.
  Oil of Turpentine 1 drm.; Olive Oil to 1 oz. *London Ophthalmic.*

Guttæ Zinci Chloridi.
  Chloride of Zinc 1 gr.; water to 1 oz. *London. London Ophthalmic. Middlesex. Westminster Ophthalmic.*
  Chloride of Zinc 2 grs.; water to 1 oz. *Royal Free.*

Guttæ Zinci Sulphatis.
  Sulphate of Zinc 2 grs.; water to 1 oz. *Gt. Northern. London Ophthalmic. Westminster. Westminster Ophthalmic.*
  Sulphate of Zinc 1 gr.; water 1 oz. *London.*
  Sulphate of Zinc 12 grs.; water 1 oz. *For the Ear. Throat.*

Guttæ Zinci Sulphatis et Aluminis.
  Sulphate of Zinc 1 gr.; Alum 3 grs.; water 1 oz. *London. London Ophthalmic.*

HAUSTUS.

Haustus Acidi Gallici.
  Glycerine of Gallic Acid 25 mins.; water to 1 oz. *St. Bartholomew's.*
  Glycerine of Gallic Acid 20 mins.; Acid Infusion of Roses 1 oz. *City Chest.*
  Gallic Acid 10 grs.; Aromatic Sulphuric Acid 15 mins.; Spirit of Chloroform 10 mins.; water to 1 oz. *Middlesex.*
  Gallic Acid 10 grs.; Dil. Sulphuric Acid 5 mins.; Acid Infusion of Roses 1 oz. *Women.*

Haustus Acidi Hydrobromici.
  Hydrobromic Acid 20 mins.; Syrup 20 mins.; water 1 oz. *City Chest.*

Haustus Acidi Hydrocyanici.
  Dil. Hydrocyanic Acid 4 mins.; Bicarbonate of Soda 5 grs.; water to 1 oz. *St. Bartholomew's.*
  Dil. Hydrocyanic Acid 6 mins.; Bicarbonate of Soda 20 grs.; water 1½ oz. *St. George's.*
  Dil. Hydrocyanic Acid 4 mins.; Bicarbonate of Soda 10 grs.; Cinnamon Water to 1 oz. *Fever.*

Haustus Acidi Hydrocyanici et Sodæ.
Bicarbonate of Soda 10 grs.; Carbonate of Magnesia 10 grs.; Dil. Hydrocyanic Acid 3 mins.; Spirit of Chloroform 10 mins.; Compound Tincture of Cardamoms 20 mins.; water to 1 oz. *Westminster.*

Haustus Acidi Nitro-hydrochlorici.
Dil. Hydrochloric Acid 15 mins.; Dil. Nitric Acid 15 mins.; Spirit of Nitrous Ether 30 mins.; Syrup 30 mins.; water to 1 oz. *Fever.*
Dil. Nitro-hydrochloric Acid 15 mins.; Compound Infusion of Gentian 1 oz. *City Chest.*
Dil. Nitro-hydrochloric Acid 10 mins.; Compound Tincture of Gentian 20 mins.; water to 1 oz. *Middlesex.*
Dil. Nitro-hydrochloric Acid 10 mins.; Tincture of Orange Peel 20 mins.; Spirit of Chloroform 10 mins.; water to 1 oz. *St. Bartholomew's. Samaritan.*
Dil. Nitro-hydrochloric Acid 10 mins.; Mint Water 1 oz. *Women.*

Haustus Acidi et Opii.
Diluted Sulphuric Acid 15 mins.; Tincture of Opium 5 mins.; Caraway Water 1 fl. oz. *Fever.*
Dil. Sulphuric Acid 15 mins.; Tincture of Opium 10 mins.; Tincture of Ginger 20 mins.; Cinnamon Water to 1 oz. *Westminster.*

Haustus Acidi Phosphorici.
Dil. Phosphoric Acid 20 mins.; water 1 oz. *Fever.*
Dil. Phosphoric Acid 20 mins.; Tincture of Orange ½ drm.; water to 1 oz. *Westminster.*

Haustus Acidi Phosphorici et Strychniæ.
Dil. Phosphoric Acid 15 mins.; Solution of Strychnia 3 mins.; Spirit of Chloroform 15 mins.; Infusion of Quassia 1 oz. *City Chest.*

Haustus Aconiti cum Colchico.
Tincture of Aconite (P.B.) 3 mins.; Tincture of Colchicum Seeds 10 mins.; Sulphate of Quinine 1 gr.; Diluted Sulphuric Acid 1½ min.; Spirit of Chloroform 10 mins.; water to 1 oz. *St. Bartholomew's.*

Haustus Ætheris Ammoniatus.
Spirit of Ether ½ drm.; Aromatic Spirit of Ammonia ½ drm.; water to 1 oz. *St. Bartholomew's.*

Haustus Ætheris et Ammoniæ.
Spirit of Ether ½ drm.; Aromatic Spirit of Ammonia ½ drm.; Camphor Water to 1 oz. *City Chest.*

Haustus Ætheris Comp.
Spirit of Ether ½ drm.; Carbonate of Ammonia 3 grs.; Spirit of Chloroform 10 mins.; water to 1 oz. *Middlesex.*
Ether 20 mins.; Dil. Sulphuric Acid 15 mins.; Peppermint Water 1 oz.; add a sufficient quantity of Sol. of Sulphate of Indigo to impart a bluish tint. *Fever.*

Haustus Ætheris et Sodæ.
Spirit of Ether 20 mins.; Bicarbonate of Soda 10 grs.; Peppermint Water 1 oz. *City Chest.*

Haustus Albus.
Sulphate of Magnesia 1 drm.; Carbonate of Magnesia 10 grs.; Peppermint Water 1 oz. *Westminster.*

Haustus Aluminis.
Alum 5 grs.; Dil. Sulphuric Acid 20 mins.; Infusion of Roses to 1 oz. *Westminster.*

Haustus Ammoniæ Acetatis.
Solution of Acetate of Ammonia 3 drms.; water to 1 oz. *Middlesex.*
Solution of Acetate of Ammonia 3 drms.; water to 1½ oz. *St. George's.*
Solution of Acetate of Ammonia 4 drms.; water to 1 oz. *St. Bartholomew's.*
Solution of Acetate of Ammonia 3 drms.; Spirit of Nitrous Ether 20 mins.; water to 1 oz. *Westminster.*

Haustus Ammoniæ Acetatis Compositus.
Strong Solution of Acetate of Ammonia 40 mins.; Carbonate of Ammonia 4 grs.; Spirit of Nitrous Ether 20 mins.; water to 1 oz. *Fever.*

Haustus Ammon. Acet. c. Camphorâ.

Solution of Acetate of Ammonia 2 drms.; Camphor water to 1 oz. *St. Bartholomew's.*

Haustus Ammoniæ Acetatis c. Ferro.

Strong Solution of Acetate of Ammonia 30 mins.; Dil. Acetic Acid 15 mins.; Tincture of Perchloride of Iron 10 mins.; water to 1 oz. *Fever.*

Solution of Acetate of Ammonia 2 drms.; Diluted Acetic Acid 10 mins.; Solution of Peracetate of Iron 15 mins.; water 1 oz. *Women.*

Solution of Acetate of Ammonia 2½ drms.; Tincture of Perchloride of Iron 10 mins.; Dil. Acetic Acid ½ drm.; water to 1 oz. *Middlesex.*

Haustus Ammon. Acet. c. Mag. Sulph.

Solution of Acetate of Ammonia 4 drms.; Sulphate of Magnesia 60 grs.; water to 1 oz. *St. Bartholomew's.*

Haustus Ammon. Acet. c. Scillâ.

Solution of Acetate of Ammonia 4 drms.; Vinegar of Squill 20 mins.; water to 1 oz. *St. Bartholomew's.*

Solution of Acetate of Ammonia 2½ drms.; Vinegar of Squill ½ drm.; water 5 drms. *City Chest.*

Haustus Ammoniæ Carbonatis.

Carbonate of Ammonia 5 grs.; water 1 oz. *Fever.*

Haustus Ammoniæ et Cinchonæ.

Carbonate of Ammonia 5 grs.; Decoction of Yellow Bark 1 oz. *Westminster.*

Haustus Ammoniæ Citratis.

Carbonate of Ammonia 16 grs.; Citric Acid 20 grs.; water to 1½ oz. *St. George's.*
Taken during effervescence.

Haustus Ammoniæ Co.

Carbonate of Ammonia 5 grs.; Aromatic Powder of Chalk 30 grs; Pimento Water 6 drms.; water to 1½ oz. *St. George's.*

Haustus Ammoniæ Effervescens.
Carbonate of Ammonia 15 grains.; water 1 oz.; Tartaric Acid 18 grs.; water ½ oz. *Fever.*

Haustus Ammoniæ et Nucis Vomicæ.
Spirit of Chloroform 15 mins.; Tincture of Nux Vomica 10 mins.; Carbonate of Ammonia 5 grs.; Infusion of Cascarilla 1 oz. *City Chest.*

Haustus Ammoniæ et Senegæ.
Carbonate of Ammonia 5 grs.; Infusion of Senega 1 oz. *Westminster.*

Haustus Ammoniaci Fœtidus.
Fœtid Spirit of Ammonia (Hosp. Pharm.) 15 mins.; Mixture of Ammoniacum to 1 oz. *St. Bartholomew's.*

Haustus Amygdalæ.
Almond Oil 2 drms.; Mucilage 2 drms.; water to 1 oz. *St. Bartholomew's.*

Haustus Antimonialis.
Antimonial Wine 1 drm.; Strong Solution of Acetate of Ammonia 30 mins.; Nitrate of Potash 5 grs.; water to 1 oz. *Fever.*

Haustus Antimonii c. Magnes Sulph.
Antimonial Wine 30 mins.; Sulphate of Magnesia 60 grs.; water to 1 oz. *St. Bartholomew's.*

Hautsus Antimonii c. Morphiâ.
Antimonial Wine 40 mins.; Solution of Hydrochlorate of Morphia 30 mins.; Camphor Water to 1 oz. *Fever.*

Haustus Aromaticus.
Aromatic Spirit of Ammonia 20 mins.; Aromatic Powder of Chalk 20 grs.; Cinnamon Water 1 oz. *City Chest.*

Haustus Aromaticus c. Rheo.
Powdered Rhubarb 3 grs.; Aromatic Powder of Chalk 10 grs.; Peppermint Water 1 oz. *Women.*

Haustus Arsenici et Ferri.
Hydrochloric Solution of Arsenic 5 mins.; Tincture of Perchloride of Iron 10 mins.; Infusion of Quassia 1 oz. *City Chest.*

Haustus Assafœtidœ.
Tincture of Assafœtida 1 drm.; water 1 oz. *Fever.*
Assafœtida 10 grs.; Mucilage 1 drm.; Camphor Water to 1 oz. *Westminster.*

Haustus Belladonnæ.
Tincture of Belladonna 20 mins.; Camphor Water to 1 oz. *Fever.*

Haustus Bismuthi.
Carbonate of Bismuth 10 grs.; Light Carbonate of Magnesia 10 grs.; Tincture of Calumba 20 mins.; water to 1 oz. *Middlesex.*
Subnitrate of Bismuth 8 grs.; Bicarbonate of Soda 10 grs.; Mucilage 1 drm.; Infusion of Quassia 1 oz. *Women.*
White Bismuth 10 grs.; Carbonate of Magnesia 15 grs.; Mucilage of Gum Arabic 2 drms.; Cinnamon Water to 1 oz. *Fever.*

Haustus Bismuthi Alkalinus.
Subnitrate of Bismuth 8 grs.; Bicarbonate of Soda 8 grs.; Infusion of Calumba 1 oz. *City Chest.*

Haustus Bismuthi et Acidi Hydrocyanici.
Subnitrate of Bismuth 10 grs.; Dil. Hydrocyanic Acid 3 mins.; Spirit of Chloroform 10 mins.; Compound Powder of Tragacanth 15 grs.; water to 1 oz. *Westminster.*

Haustus Bismuthi Compositus.
Subnitrate of Bismuth 10 grs.; Bicarbonate of Soda 15 grs.; Diluted Hydrocyanic Acid 3 mins.; Mucilage 1 drm.; Infusion of Calumba 1 oz. *Women.*

Haustus Bismuthi et Gentianæ.
Subnitrate of Bismuth 10 grs.; Compound Powder of Tragacanth 15 grs.; Compound Infusion of Gentian 1 oz. *Westminster.*

Haustus Calcis Hypophosphitis.
Hypophosphite of Lime 3 grs.; Saccharated Solution of Lime 10 mins.; Glycerine 20 mins.; Camphor Water 1 oz. *City Chest.*

Haustus Calcis Hypophosphitis Comp.
Hypophosphite of Lime 5 grs.; Syrup of Phosphate of Iron 60 mins.; Glycerine 60 mins.; water to 1 oz. *Middlesex.*

Haustus Calumbæ Alkalinus.
Bicarbonate of Soda 10 grs.; Tincture of Calumba 20 mins.; Infusion of Orange Peel to 1 oz. *St. Bartholomew's*

Haustus Calumbæ et Ammoniæ.
Carbonate of Ammonia 5 grs.; Infusion of Calumba 1 oz. *City Chest.*

Haustus Calumbæ et Sodæ.
Bicarbonate of Soda 10 grs.; Infusion of Calumba 1 oz. *City Chest.*

Haustus Camphoræ Fœtidus.
Fœtid Spirit of Ammonia 15 mins.; Camphor Water to 1 oz. *Samaritan.*

Haustus Cannabis Indicæ.
Extract of Indian Hemp 1 gr.; Ether 15 mins. Camphor Water 1 oz. *Fever.*
Tincture of Indian Hemp 20 mins.; Syrup of Chloral 20 mins; Glycerine 1 drm.; Water 1 oz. *Women.*

Haustus Carminativus.
Rhubarb 5 grs.; Ginger 5 grs.; Bicarbonate of Soda 10 grs.; Aromatic Spirit of Ammonia 20 mins.; Cinnamon Water to 1 oz. *Westminster.*

Haustus Caryophylli Co.
Bicarbonate of Soda 10 grs.; Spirit of Chloroform 10 mins.; Aromatic Spirit of Ammonia 20 mins.; Infusion of Cloves to 1 oz. *Middlesex.*

Haustus Cascarillæ.
Compound Tincture of Camphor 20 mins.; Spirit of Chloroform 10 mins.; Oxymel of Squill ½ drm.; Infusion of Cascarilla to 1 oz. *Westminster.*

Haustus Cascarillæ Compositus.
Vinegar of Squill 20 mins.; Compound Tincture of Camphor 20 mins.; Infusion of Cascarilla to 1 oz. *City Chest.*

Haustus Cascarillæ c. Scillâ.
Tincture of Squill 15 mins.; Infusion of Cascarilla to 1 oz. *St. Bartholomew's.*

Haustus Catharticus.
Senna 1 oz.; Ginger 2 drms.; boiling water 20 oz. Infuse for 1 hour, strain, and add Sulphate of Magnesia 4 oz.; Sal Volatile 1 oz. Dose 1 to 2 oz. *Gt. Northern.*

Haustus Cetacei.
Spermaceti 20 grs.; Powd. Acacia Gum 20 grs.; water 6 drms.; Pimento Water to 1½ oz. *St. George's.*

Haustus Cetacei Co.
Spermaceti 20 grs.; Powd. Tragacanth 5 grs.; Syrup of Poppies 30 mins.; water to 1 oz. *Middlesex.*

Comp. Tincture of Camphor 1 drm.; Vinegar of Squill ½ drm.; Haustus Cetacei to 1½ oz. *St. George's*

Haustus Chalybeatus.
Tincture of Perchloride of Iron 15 mins.; Dil. Nitric Acid 15 mins.; Syrup 20 mins.; Peppermint Water to 1 oz. *City Chest.*

Haustus Chiratæ.
Infusion of Chiretta 1 oz. *Fever.*

Haustus Chlori.
Chlorate of Potash ½ gr.; Hydrochloric Acid 3 mins.; mix for 5 minutes, and add water 7½ drms.; Chlorate of Potash 10 grs.; Spirit of Chloroform 10 mins.; Dil. Hydrochloric Acid 15 mins. *Fever.*

Haustus Chloroformi Compositus.

Spirit of Chloroform 15 mins.; Comp. Tincture of Camphor 20 mins.; Syrup of Squill ½ drm.; water to 1 oz. *Middlesex.*

Chloroform 1 min.; Diluted Hydrocyanic Acid 3 mins.; Solution of Acetate of Morphia 3 mins.; Diluted Nitric Acid 10 mins.; Glycerine 1 drm.; Tincture of Cascarilla ½ drm.; water to 1 oz. *St. Bartholomew's.*

Haustus Cinchonæ.

Tincture of Yellow Bark 1½ drm.; Decoction of Yellow Bark to 1½ oz. *St. George's.*

Dil. Nitro-hydrochloric Acid 10 mins.; Spirit of Chloroform 5 mins.; Decoction of Bark to 1 oz. *Middlesex.*

Haustus Cinchonæ Acidus.

Aromatic Sulphuric Acid 10 mins.; Spirit of Chloroform 10 mins.; Decoction of Pale Bark to 1 oz. *St. Bartholomew's.*

Dil. Hydrochloric Acid 10 mins.; Decoction of Yellow Bark 1 oz. *Westminster.*

Dil. Nitro-hydrochloric Acid 10 mins.; Decoction of Yellow Bark 1 oz. *Women.*

Dil. Sulphuric Acid 5 mins.; Decoction of Bark 1 oz. *City Chest.*

Dil. Sulphuric Acid 15 mins.; Decoction of Yellow Bark 1½ oz. *St. George's.*

Haustus Cinchonæ c. Ammoniâ.

Carbonate of Ammonia 5 grs.; Spirit of Chloroform 5 mins.; Decoction of Bark to 1 oz. *Middlesex.*

Haustus Cinchoniæ.

Hydrochlorate of Cinchonine 1 gr.; Dil. Hydrochloric Acid 10 mins.; Treacle ½ drm.; water to 1 oz. *Fever.*

Haustus Cinchoninæ c. Ferro.

Sulphate of Cinchonine 1 gr.; Sulphate of Iron 1 gr.; Dil. Sulphuric Acid 2 mins.; water to 1 oz. *Middlesex.*

Haustus Cinchonæ Hydrochloratis.
Hydrochlorate of Cinchonine 1 gr.; Diluted Hydrochloric Acid 1½ min.; water to 1 oz. *St. Bartholomew's.*

Haustus Colchici.
Bicarbonate of Potash 15 grs.; Tincture of Colchicum Seeds 20 mins.; Tincture of Hyoscyamus 20 mins.; Spearmint Water 1 oz. *Westminster.*

Tincture of Colchicum Seeds 15 mins.; Carbonate of Magnesia 10 grs.; Spearmint Water to 1 oz. *St. Bartholomew's.*

Haustus Colchici et Opii.
Wine of Colchicum 15 mins.; Solution of Acetate of Ammonia (strong) 40 mins.; Dover's Powder 5 grs.; Camphor Julep 1 oz. *St. George's.*

Haustus Copaibæ.
Copaiba 30 mins.; Solution of Potash 20 mins.; Mucilage of Acacia 2 drms.; Caraway Water to 1 oz. *Middlesex.*

Copaiba 15 mins.; Solution of Potash 15 mins.; Spirit of Nitrous Ether 30 mins.; Camphor Water to 1 oz. *City Chest.*

Haustus Copaibæ Co.
Copaiba 15 mins.; Cubebs 30 grs.; Spirit of Nitrous Ether 15 mins.; Mucilage of Acacia 1 drm.; Camphor Water to 1 oz. *St. Bartholomew's.*

Haustus Copaibæ Resinæ.
Tincture of Copaiba (3 mins. contain 1 grain) 30 mins.; Spirit of Chloroform 10 mins.; Syrup of Ginger 60 mins.; Mucilage of Acacia 60 mins.; water to 1 oz. *Middlesex. Westminster.*

Haustus Cretæ Aromaticus.
Aromatic Powder of Chalk 30 grs.; Cassia Water to 1 oz. *St. Bartholomew's.*

Aromatic Powder of Chalk 30 grs.; Cinnamon Water 1 oz. *Westminster.*

Haustus Cretæ Aromat. c. Opio.
Aromatic Powder of Chalk and Opium 20 grs.; Cassia Water to 1 oz. *St. Bartholomew's.*

Haustus Cretæ Aromat. c. Rheo.
: Aromatic Powder of Chalk 20 grs. ; Powder of Rhubarb 10 grs. ; Cassia Water 1 oz. *St. Bartholomew's.*

Haustus Cretæ c. Catechu.
: Tincture of Catechu 1 drm. ; Chalk Mixture to 1 oz. *Fever.*

Haustus Cretæ Compositus.
: Chalk Mixture 1 oz. ; Tincture of Catechu 1 drm. ; Glycerine of Carbolic Acid 5 mins. *Women.*
: Prepared Chalk 15 grs.; Mucilage of Acacia 60 mins.; Tincture of Catechu 30 mins. ; water to 1 oz. *Middlesex.*

Haustus Cretæ et Opii.
: Tincture of Opium 10 mins. ; Tincture of Catechu ½ drm. ; Chalk Mixture to 1 oz. *Westminster.*
: Tincture of Opium 5 mins. ; Tincture of Catechu ½ drm.; Chalk Mixture 1 oz. *City Chest.*

Haustus Diaphoreticus.
: Tartarated Antimony $\frac{1}{12}$ gr. ; Solution of Acetate of Ammonia 2 drms.; water to 1 oz. *Westminster.*

Haustus Diureticus.
: Acid Tartrate of Potash 40 grs. ; Tincture of Digitalis 10 mins. ; Spirit of Nitrous Ether 30 mins. ; water to 1 oz. *Fever.*
: Acetate of Potash 20 grs. ; Spirit of Nitrous Ether ½ drm.; Decoction of Broom to 1 oz. *Westminster.*

Haustus Effervescens.
: Bicarbonate of Soda 20 grs.; water 1 oz.; Tartaric Acid 17 grs.; water ½ oz. *City Chest. Fever. Westminster.*

Haustus Effervescens c. Acido Hydrocyanico.
: Bicarbonate of Soda 20 grs. ; water 1 oz. ; Tartaric Acid 18 grs. ; Dil. Hydrocyanic Acid 4 mins.; water 1 oz. *Fever.*

Haustus Effervescens c. Calumbâ.
: Bicarbonate of Soda 20 grs. ; Infusion of Calumba 1 oz.; Tartaric Acid 17 grs.; Infusion of Calumba ½ oz. *City Chest.*

Haustus Emeticus.
Wine of Ipecacuanha 4 drms. ; Oxymel of Squill 4 drms. *St. Bartholomew's.*
Antimonial Wine 2 drms. ; Wine of Ipecacuanha 6 drms. *Fever.*
Ipecacuanha 10 grs. ; Sulphate of Zinc 20 grs. ; Syrup 1 drm. ; Spearmint Water to 1 oz. *Westminster.*

Haustus Emeticus Purgans.
Tartarated Antimony $\frac{1}{3}$ gr. ; Sulphate of Magnesia 60 grs. ; water to 1 oz. *Middlesex.*

Haustus Ergotæ Acidus.
Dil. Sulphuric Acid 10 mins.; Infusion of Ergot to 1 oz. *Westminster.*

Haustus Ergotæ Compositus.
Carbonate of Ammonia 4 grs.; Liquid Extract of Ergot 15 to 30 mins.; Comp. Tincture of Lavender 15 mins.; Spirit of Chloroform 10 mins.; Camphor Water 1 oz. *Women.*

Haustus Ferri Acidus.
Tincture of Perchloride of Iron 15 mins. ; water to 1 oz. *Fever.*

Haustus Ferri Alkalinus.
Tartarated Iron 10 grs.; Bicarbonate of Potash 10 grs. ; water 1 oz. *Westminster.*
Ammonio-Citrate of Iron 5 grs.; Bicarbonate of Soda 15 grs. ; Infusion of Calumba 1 oz. *City Chest.*

Haustus Ferri c. Magnesiâ.
Sulphate of Iron 1 gr.; Sulphate of Magnesia 20 grs.; Diluted Sulphuric Acid 10 mins. ; Peppermint Water to 1 oz. *Middlesex.*

Haustus Ferri et Ammoniæ.
Citrate of Iron and Ammonia 5 grs. ; water to 1 oz. *Fever.*
Citrate of Iron and Ammonia 8 grs.; Carbonate of Ammonia 2 grs.; Spirit of Chloroform 10 mins.; Infusion of Quassia to 1 oz. *Samaritan.*

Haustus Ferri et Ammoniæ Citratis.

Citrate of Iron and Ammonia 8 grs. ; Carbonate of Ammonia 2 grs. ; Spirit of Chloroform 10 mins. ; water to 1 oz. *St. Bartholomew's.*

Citrate of Iron and Ammonia 5 grs.; Spirit of Chloroform 5 mins. ; Infusion of Quassia to 1 oz. *Middlesex.*

Haustus Ferri Aperiens.

Sulphate of Iron 2 grs. ; Sulphate of Magnesia 40 grs.; Diluted Sulphuric Acid 5 mins.; Peppermint Water 1 oz. *Women.*

Sulphate of Iron 2 grs.; Sulphate of Magnesia 30 grs. ; Dil. Sulphuric Acid 10 mins.; Spirit of Chloroform 10 mins.; Peppermint Water to 1 oz. *Westminster.*

Haustus Ferri Citratis.

Ammonio-Citrate of Iron 10 grs.; Pimento Water 1 oz. *Westminster*

Haustus Ferri Iodidi.

Tartarated Iron 10 grs. ; Iodide of Potassium 5 grs. ; water 1 oz. *Westminster.*

Haustus Ferri cum Nuce Vomicâ.

Solution of Perchloride of Iron 10 mins.; Tincture of Nux Vomica 6 mins. ; Diluted Phosphoric Acid 5 mins. ; Spirit of Chloroform 5 mins.; water to 1 oz. *St. Bartholomew's.*

Haustus Ferri Perchloridi.

Solution of Perchloride of Iron 10 mins. ; Glycerine 15 mins. ; water 1 oz. *Women.*

Tincture of Perchloride of Iron 15 mins.; Spirit of Chloroform 10 mins.; water to 1 oz. *Middlesex.*

Tincture of Perchloride of Iron 15 mins. ; Spirit of Chloroform 10 mins. ; Glycerine 20 mins.; water to 1 oz. *Westminster.*

Haustus Ferri Phosphatis.

Mistura Ferri Compositus (P.B.) cum Sodæ Phosphate vice Potassæ Carbonate. *City Chest.*

Haustus Ferri et Quassiæ.
Sulphate of Iron 1 gr.; Dil. Sulphuric Acid 5 mins.; Infusion of Quassia 1 oz. *City Chest.*
Solution of Perchloride of Iron 15 mins.; Infusion of Quassia to 1 oz. *St. Bartholomew's.*

Haustus Ferri et Quiniæ.
Sulphate of Quinia 2 grs.; Tincture of Perchloride of Iron 20 mins.; Spirit of Chloroform 15 mins.; water to 1 oz. *Fever.*

Haustus Ferri et Quiniæ Citratis.
Citrate of Iron and Quinine 5 grs.; Aromatic Spirit of Ammonia ½ drm.; water 1 oz. *City Chest.*

Haustus Filicis. (*See also* Mist. Filicis.)
Liquid Extract of Male Fern 1½ drm.; Mucilage 2 drms.; Peppermint Water to 1½ oz. *Westminster.*
Liquid Extract of Male Fern 1 drm.; Syrup of Ginger 1 drm.; Tincture of Quillaia ½ drm.; water to 1½ oz. *London.*
Liquid Extract of Male Fern 1 drm.; Glycerine 1 drm.; Mucilage 1 drm.; Syrup of Ginger 1 drm.; water to 1 oz. *Middlesex.*
Liquid Extract of Male Fern 1 drm.; Mucilage 4 drms.; water to 1 oz. *Royal Free.*

Haustus Flavus.
Caramel a sufficiency; water to 1 oz. *St. Bartholomew's.*

Haustus Gentianæ Acidus.
Dil. Hydrochloric Acid 10 mins.; Compound Infusion of Gentian to 1 oz. *Westminster.*
Aromatic Sulphuric Acid 10 mins.; Simple Infusion of Gentian to 1 oz. *Middlesex.*

Haustus Gentianæ Alkalinus.
Bicarbonate of Soda 15 grs.; Compound Infusion of Gentian 1 oz. *Westminster.*
Bicarbonate of Potash 10 grs.; Tincture of Ginger 10 mins.; Infusion of Rhubarb 60 mins.; Simple Infusion of Gentian to 1 oz. *Middlesex.*

Haustus Gentianæ c. Rheo.
Compound Tincture of Gentian ½ drm. ; Bicarbonate of Soda 10 grs. ; Spirit of Chloroform 10 mins. ; Infusion of Rhubarb ½ oz. ; Peppermint Water to 1 oz. *St. Bartholomew's. Samaritan.*

Haustus Gentianæ c. Sennâ.
Infusion of Senna ½ oz. ; Tincture of Ginger 20 mins. ; Comp. Infusion of Gentian to 1½ oz. *Westminster.*
Infusion of Senna 2½ drms. ; Compound Infusion of Gentian 5½ drms. *City Chest.*

Haustus Gentianæ c. Sodâ.
Bicarbonate of Soda 10 grs. ; Compound Infusion of Gentian 1 oz. *Women.*

Haustus Guaiaci Co.
Ammoniated Tincture of Guaiacum 40 mins. ; Mucilage 2 drms. ; water to 1 oz. *Middlesex.*

Haustus Hemidesmi c. Potassii Iodido.
Comp. Decoction of Hemidesmus 1 oz. ; Iodide of Potassium 5 grs. *St. Bartholomew's.*

Haustus Hydrargyri Biniodidi.
Red Iodide of Mercury $\frac{1}{20}$ gr. ; Iodide of Potassium 5 grs. ; Decoction of Yellow Bark 1 oz. *Westminster.*

Haustus Hydrargyri Iodidi.
Solution of Perchloride of Mercury 60 mins.; Iodide of Potassium 5 grs. ; Tincture of Yellow Bark 30 mins. ; water to 1 oz. *Middlesex.*

Haustus Hydrargyri Perchloridi.
Solution of Perchloride of Mercury 60 mins ; Syrup of Red Poppies 30 mins; water to 1 oz. *St. Bartholomew's.*

Haustus Hydrargyri Perchloridi cum Potassii Iodido.
Solution of Perchloride of Mercury 60 mins.; Iodide of Potassium 5 grs. ; Compound Tincture of Cardamoms 20 mins. ; water to 1 oz. *St. Bartholomew's.*

Haustus Hyoscyami.
Tincture of Henbane 1 drm.; Camphor Water to 1 oz. *Fever.*

Haustus Imperialis.
Acid Tartrate of Potash 1 drm.; Sugar 4 drms.; boiling water 20 oz. *University.*

Haustus Ipecacuanhæ.
Solution of Potash 2 mins.; Wine of Ipecacuanha 2½ mins.; Syrup 20 mins.; Mucilage 30 mins.; Dill Water to 2 drms. *Westminster.*

Haustus Ipecacuanhæ c. Ammoniâ.
Carbonate of Ammonia 4 grs.; Ipecacuanha Wine 10 mins.; Spirit of Chloroform 10 mins; Syrup of Extract of Poppies 30 mins; water to 1 oz. *Middlesex.*

Haustus Ipecacuanhæ et Ammoniæ.
Carbonate of Ammonia 10 grs.; Ipecacuanha Wine 15 mins.; Chlorate of Potash 20 grs.; water 1 oz. *Fever.*

Haustus Ipecacuanhæ Co.
Ipecacuanha Wine 10 mins.; Comp. Tincture of Camphor 30 mins.; Solution of Acetate of Ammonia 3 drms.; water to 1 oz. *Middlesex.*

Haustus Ipecacuanhæ Emeticus.
See Haustus Emeticus.

Haustus Ipecacuanhæ Opiatus.
Ipecacuanha Wine 8 mins.; Tincture of Opium 4 mins.; Nitrate of Potash 8 grs.; Mucilage 2 drms.; water 6 drms. *City Chest.*

Haustus Ipecacuanhæ c. Scillâ.
Wine of Ipecacuanha 10 mins; Syrup of Squill 30 mins.; Spearmint Water to 1 oz. *Middlesex.*

Haustus Magnesiæ Acidus.
Sulphate of Magnesia 1 drm.; Dil. Sulphuric Acid 10 mins.; Syrup of Red Poppies 30 mins.; Spearmint Water to 1 oz. *Westminster.*

Haustus Magnes. Co. Vel Haustus Albus.

Carbonate of Magnesia 15 grs.; Aromatic Spirit of Ammonia ½ drm.; Compound Tincture of Camphor 15 mins.; Dill Water 1 oz. *City Chest.*

Carbonate of Magnesia 20 grs.; Sulphate of Magnesia 60 grs.; Peppermint Water to 1 fl. oz. *Consumption.*

Light Carbonate of Magnesia 5 grs.; Sulphate of Magnesia 60 grs.; Peppermint Water to 1 oz. *Middlesex.*

Haustus Magnesiæ et Magnesiæ Sulphatis.

Carbonate of Magnesia 15 grs.; Sulphate of Magnesia ½ drm.; Peppermint Water 1 oz. *City Chest.*

Haustus Magnesiæ c. Rheo.

Light Carbonate of Magnesia 15 grs.; Sulphate of Magnesia 60 grs.; Powd. Rhubarb 20 grs.; Peppermint Water to 1 oz. *Middlesex.*

Carbonate of Magnesia 5 grs.; Sulphate of Magnesia 60 grs.; Powd. Rhubarb 20 grs.; Tinct. of Ginger 5 mins.; Peppermint Water 1 oz. *Women.*

Haustus Magnesiæ c. Sodâ.

Carbonate of Magnesia 8 grs.; Bicarbonate of Soda 8 grs.; Caraway Water 1 oz. *St. Bartholomew's.*

Haustus Menthæ Sulphuricus.

Dil. Sulphuric Acid 10 mins.; Syrup ½ drm.; Peppermint Water 1 oz. *City Chest.*

Dil. Sulphuric Acid 10 mins.; Syrup of Red Poppies 30 mins.; Spearmint Water to 1 oz. *St. Bartholomew's.*

Haustus Menth. Sulph. c. Ferri Sulph.

Sulphate of Iron 1 gr.; Spirit of Chloroform 10 mins.; Sulphuric Mint Draught to 1 oz. *St. Bartholomew's.*

Haustus Menthæ Sulph. c. Magn. Sulph.

Dil. Sulphuric Acid 10 mins.; Sulphate of Magnesia 60 grs.; Syrup of Red Poppies 30 mins.; Spearmint Water to 1 oz. *St. Bartholomew's. Samaritan.*

Haustus Menthæ Sulphuricus c. Papavere et Scillâ.
> Syrup of Poppies ½ drm.; Tincture of Squill 15 mins.; Sulphuric Mint Draught to 1 oz. *St. Bartholomew's.*

Haustus Morphiæ.
> Hydrochlorate of Morphia ⅙ gr.; Dil. Hydrochloric Acid 1 min.; water to 1 oz. *Westminster.*

Haustus Morphiæ Acetatis.
> Acetate of Morphia ¼ gr.; Pimento Water 6 drms. water to 1½ oz. *St. George's.*

Haustus Morphiæ et Acidi Hydrocyanici.
> Solution of Hydrochlorate of Morphia 10 mins.; Dil. Hydrocyanic Acid 3 mins.; Spirit of Chloroform 10 mins.; water to 1 oz. *Westminster.*

Haustus Morphiæ c. Ferro.
> Solution of Hydrochlorate of Morphia 10 mins.; Solution of Perchloride of Iron 15 mins.; Dil. Hydrochloric Acid 3 mins.; Spirit of Chloroform 10 mins.; water to 1 oz. *St. Bartholomew's. Samaritan.*
> Solution of Hydrochlorate of Morphia 10 mins.; Tincture of Perchloride of Iron 15 mins.; Spirit of Chloroform 10 mins.; water 1 oz. *City Chest.*

Haustus Morphiæ Hydrochloratis.
> Solution of Hydrochlorate of Morphia 20 mins.; water to 1 oz. *St. Bartholomew's.*

Haustus Niger. *See* Mistura Sennæ.

Haustus Nitrosus.
> Dil. Nitric Acid 10 mins.; Spirit of Nitrous Ether 20 mins.; Syrup 30 mins.; Camphor Water to 1 oz. *City Chest.*

Haustus Olei Morrhuæ.
> Cod Liver Oil 2 drms.; Mucilage of Acacia 2 drms.; White Sugar 30 grs.; Caraway Water to 1 oz. *St. Bartholomew's.*

Haustus Olei Morrhuæ Alkalinus.
> Cod Liver Oil 2 drms.; Solution of Potash 15 mins.; Caraway Water to 1 oz. *Westminster.*

Haustus Olei Ricini.
Castor Oil ½ oz.; Solution of Potash 15 mins.; Peppermint Water to 1 oz. *Middlesex.*

Haustus Olei Ricini Alkalinus.
Castor Oil 2 drms.; Solution of Potash 15 mins.; Caraway Water to 1 oz. *Westminster.*

Haustus Olei Ricini c. Terebinthinâ.
Castor Oil 1 oz.; Oil of Turpentine 2 drms.; Peppermint Water to 2 oz. *Royal Free.*

Haustus Opiatus.
Tincture of Opium 15 mins.; Pimento Water 6 drms.; water to 1½ oz. *St. George's.*
Tincture of Opium 10 mins.; Spirit of Chloroform 10 mins.; Peppermint Water to 1 oz. *St. Bartholomew's.*

Haustus Opii c. Æthere.
Tincture of Opium 15 mins.; Ether 15 mins.; Camphor Water to 1 oz. *Fever.*

Haustus Pimentæ.
Pimento Water 6 drms.; water to 1½ oz. *St. George's.*

Haustus Plumbi.
Acetate of Lead 3 grs.; Dil. Acetic Acid 5 mins.; Distilled Water to 1 oz. *Fever.*

Haustus Plumbi Acetatis.
Acetate of Lead 5 grs.; Oxymel 1 drm.; water to 1 oz. *St. Bartholomew's.*

Haustus Plumbi c. Morphiâ.
Acetate of Lead 3 grs.; Acetate of Morphia ⅙ gr.; Dil. Acetic Acid 5 mins.; Peppermint Water to 1 oz. *Fever.*

Haustus Potassæ Acetatis Comp.
Acetate of Potash 15 grs.; Tincture of Squill 15 mins.; Sweet Spirit of Nitre 20 mins.; Decoction of Broom 1 oz.; mix. *Women.*

Haustus Potassæ Acetatis et Scillæ.
Acetate of Potash 20 grs.; Spirit of Nitrous Ether ½ drm.; Solution of Acetate of Ammonia 2 drms.; Vinegar of Squill 20 mins.; water to 1 oz. *City Chest.*

Haustus Potassæ Bicarbonatis.
Bicarbonate of Potash 30 grs.; Nitrate of Potash 10 grs.; water to 1 oz. *Fever.*

Haustus Potassæ Chloratis.
Chlorate of Potash 10 grs.; Boiling Water to 1 oz. *St. Bartholomew's.*
Chlorate of Potash 20 grs.; water to 1 oz. *Fever.*

Haustus Potassæ Chloratis Acidus.
Chlorate of Potash 10 grs.; Dil. Hydrochloric Acid 10 mins.; Decoction of Bark 1 oz. *City Chest.*

Haustus Potassæ Chloratis c. Ferro.
Chlorate of Potash 20 grs.; Tincture of Perchloride of Iron 20 mins.; water to 1 oz. *Fever.*
Chlorate of Potash 10 grs.; Tincture of Perchloride of Iron 10 mins.; Dil. Hydrochloric Acid 5 mins.; water to 1 oz. *Middlesex.*

Haustus Potassæ Citratis.
Bicarbonate of Potash 20 grs.; Citric Acid 14 grs.; water to 1 oz. *Middlesex. St. George's. St. Bartholomew's.*
Citrate of Potash 20 grs.; Syrup ½ drm.; water to 1 oz. *Westminster.*

Haustus Potassæ Citratis Effervescens.
Bicarbonate of Potash 20 grs.; Citric Acid 14 grs.; water to 1 oz. Taken during effervescence. *Middlesex. St. George's.*
Bicarbonate of Potash 20 grs.; water 1 oz.; Citric Acid 14 grs.; water 1 oz. Taken during effervescence. *St. Bartholomew's.*

Haustus Potassæ c. Hyoscyamo.
Nitrate of Potash 10 grs.; Bicarbonate of Potash 20 grs.; Tincture of Hyoscyamus 20 mins.; water to 1 oz. *St. Bartholomew's.*

Haustus Potassæ c. Magnesiæ Sulphate.
Nitrate of Potash 10 grs.; Bicarbonate of Potash 20 grs.; Tartarated Soda* 60 grs.; Peppermint Water to 1 oz. *St. Bartholomew's.*

---

* This draught may be employed either with Sulphate of Magnesia or Tartarated Soda as desired, it keeps best with the latter.

Haustus Potassæ Nitratis.
: Nitrate of Potash 10 grs.; Mint Water ½ oz.; water to 1½ oz. *St. George's.*

Haustus Potassæ Tart. v. Haust. Salinus.
: Bicarbonate of Potash 20 grs.; Spearmint Water ½ oz.; Tartaric Acid 15 grs.; water 1 oz. *St. George's.*

Haustus Potassii Bromidi.
: Bromide of Potassium 20 grs.; Aromatic Spirit of Ammonia 20 mins.; water to 1 oz. *Westminster.*
: Bromide of Potassium 20 grs.; Infusion of Quassia 1 oz. *Middlesex.*

Haustus Potassii Bromidi Compositus.
: Bromide of Potassium 10 grs.; Infusion of Rhubarb ½ oz.; Compound Tincture of Gentian 30 mins.; Spirit of Chloroform 10 mins.; Peppermint Water to 1 oz. *St. Bartholomew's.*

Haustus Potassii Bromidi c. Ferro.
: Bromide of Potassium 10 grs.; Carbonate of Ammonia 3 grs.; Ammonio-Citrate of Iron 4 grs.; Spirit of Chloroform 10 mins.; Water 1 oz. *Women.*

Haustus Potassii Bromidi et Iodidi.
: Bromide of Potassium 5 grs.; Iodide of Potassium 3 grs.; Compound Infusion of Gentian 1 oz. *Westminster.*

Haustus Potassii Iodidi.
: Iodide of Potassium 3 grs; Carbonate of Potash 10 grs.; Pimento Water 6 drms.; water to 1½ oz. *St. George's.*
: Iodide of Potassium 5 grs.; Aromatic Spirit of Ammonia 20 mins.; Decoction of Cinchona to ·1 oz. *Middlesex.*
: Iodide of Potassium 5 grs.; Aromatic Spirit of Ammonia 20 mins.; water to 1 oz. *Westminster.*

Haustus Potassii Iodidi Alkalinus.
: Iodide of Potassium 5 grs.; Bicarbonate of Potash 15 grs; Tincture of Hyoscyamus 20 mins.; Spirit of Chloroform 10 mins.; water to 1 oz. *Westminster.*
: Iodide of Potassium 2 grs.; Bicarbonate of Potash 15 grs.; Simple Infusion of Gentian 1 oz.; *Middlesex.*

Haustus Potassii Iodidi Ammoniatus.
Iodide of Potassium 3 grs.; Carbonate of Ammonia 2 grs.; water 1 oz. *St. Bartholomew's.*

Haustus Potassii Iodidi c. Colchico.
Iodide of Potassium 5 grs.; Bicarbonate of Potash 10 grs.; Colchicum Wine 10 mins.; Simple Infusion of Gentian to 1 oz. *Middlesex.*

Haustus Potassii Iodidi Compositus.
Iodide of Potassium 3 grs.; Tincture of Opium 3 mins.; Aromatic Spirit of Ammonia ½ drm.; Decoction of Bark 1 oz. *City Chest.*

Haustus Quassiæ.
Tincture of Ginger 1 drm.; Infusion of Quassia to 1½ oz. *St. George's.*

Haustus Quassiæ Acidus.
Diluted Nitric Acid 10 mins.; Infusion of Quassia 1 oz. *Women.*

Haustus Quassiæ Alkalinus.
Bicarbonate of Soda 15 grs.; Sal Volatile 15 mins.; Infusion of Quassia 1 oz. *Women.*

Haustus Quassiæ et Ferri.
Tincture of Sesquichloride of Iron 15 mins.; Quassia Draught 1½ oz.. *St. George's.*
Tincture of Perchloride of Iron 15 mins.; Infusion of Quassia to 1 oz. *Westminster.*

Haustus Quiniæ.
Sulphate of Quinia 1 gr.; Dil. Sulphuric Acid 1 min.; Water 1 oz. *Middlesex.*
Sulphate of Quinia 1 gr.; Dil. Sulphuric Acid 5 mins.; Infusion of Quassia 1 oz. *City Chest.*
Sulphate of Quinia 1 gr.; Dil. Sulphuric Acid 1½ mins.; water 1 oz. *St. Bartholomew's.*
Sulphate of Quinia 2 grs.; Dil. Sulphuric Acid 10 mins.; water 1 oz. *Fever.*
Sulphate of Quinia 2 grs.; Dil. Sulphuric Acid 3 mins.; water to 1 oz. *Westminster.*
Sulphate of Quinia 2 grs.; Dil. Sulphuric Acid 4 mins.; Tincture of Orange 1 drm.; water to 1½ oz. *St. George's.*

Haustus Quiniæ et Acidi Hydrochlorici.
Sulphate of Quinia 1 gr.; Dil. Hydrochloric Acid 10 mins.; Infusion of Quassia 1 oz. *City Chest.*

Haustus Quiniæ Acidus.
Sulphate of Quinia 2 grs.; Dil. Sulphuric Acid ½ drm.; Comp. Tincture of Cardamoms 1 drm.; Syrup 1 drm.; water to 1½ oz. *St. George's.*

Haustus Quiniæ Ammoniatus.
Ammoniated Tincture of Quinia 1 drm.; Spirit of Chloroform 10 mins.; water to 1 oz. *Middlesex.*

Haustus Quiniæ et Digitalis.
Sulphate of Quinia 2 grs.; Tincture of Digitalis 10 mins.; Nitrate of Potash 10 grs.; water 1 oz. *City Chest.*

Haustus Quiniæ c. Ferro.
Sulphate of Quinia 1 gr.; Sulphate of Iron 1 gr.; Dil. Sulphuric Acid 5 mins.; Infusion of Quassia 1 oz. *City Chest.*
Sulphate of Quinia 2 grs.; Sulphate of Iron 2 grs.; Dil. Sulphuric Acid 10 mins.; water 1½ oz. *St. George's.*
Sulphate of Quinia 1 gr.; Sulphate of Iron 1 gr.; Dil. Sulphuric Acid 2 mins.; water to 1 oz. *Middlesex.*
Sulphate of Quinia 1 gr.; Sulphate of Iron 2 grs.; Dil. Sulphuric Acid 4 mins.; water to 1 oz. *Westminster.*
Sulphate of Quinia 1 gr.; Sulphate of Iron ½ gr.; Spirit of Chloroform 10 mins.; Sulphuric Mint Draught to 1 oz. *St. Bartholomew's.*

Haustus Quiniæ c. Opio.
Sulphate of Quinia 1 gr.; Dil. Sulphuric Acid 15 mins.; Tincture of Opium 5 mins.; Peppermint Water 1 oz. *Fever.*

Haustus Quiniæ c. Potassii Iodido.
Sulphate of Quinia 1 gr.; Dil. Sulphuric Acid 1½ mins.; Iodide of Potassium 5 grs.; water to 1 oz. *St. Bartholomew's.*

Haustus Quinquiniæ.
Quinquiniæ Hydrochloras 1 gr.; Dil. Phosphoric Acid 5 mins.; Chloroform Water 1 oz. *Women.*

Haustus Quinquiniæ c. Ferro.
Quinquiniæ Sulphatis 1 gr.; Sulphate of Iron 1 gr.; Dil. Sulphuric Acid 3 mins.; Glycerine 20 mins.: water 1 oz. *Women.*

Haustus Rhei.
Rhubarb 20 grs.; Aromatic Powder of Chalk 20 grs.; Pimento Water 6 drms.; water to 1½ oz. *St. George's.*

Haustus Rhei Aromaticus.
Tincture of Rhubarb 1 drm.; Aromatic Powder of Chalk 10 grs.; Cinnamon Water 7 drms. *City Chest.*

Haustus Rhei Co.
Rhubarb 20 grs.; Carbonate of Magnesia 20 grs.; Aromatic Spirit of Ammonia 20 mins.; Pimento Water 6 drms.; water 1½ oz. *St. George's.*

Haustus Rhei c. Magnesiâ.
Rhubarb 12 grs.; Carbonate of Magnesia 12 grs.; Aromatic Spirit of Ammonia ½ drm.; Peppermint Water 12 drms. *City Chest.*
Comp. Rhubarb Powder 30 grs.; Caraway Water to 1 oz. *St. Bartholomew's.*
Rhubarb 15 grs.; Carbonate of Magnesia 30 grs.; Aromatic Spirit of Ammonia 30 mins.; Pimento Water to 1½ oz. *Consumption.*

Haustus Rhei c. Sodâ.
Rhubarb 10 grs.; Bicarbonate of Soda 15 grs.; Tincture of Ginger 10 mins.; Stronger Solution of Ammonia 2 mins.; Peppermint Water to 1 oz. *Consumption.*

Haustus Rosæ Aperiens.
Sulphate of Magnesia ½ drm.; Acid Infusion of Roses 1 oz. *City Chest.*

Haustus Salinus.
Citrate of Potash 20 grs.; water 1 oz. *Women.*
Nitrate of Potash 10 grs.; Spirit of Nitrous Ether ½ drm.; Solution of Acetate of Ammonia 2 drms.; water to 1 oz. *City Chest.*

Haustus Salinus Antimonialis.
Antimonial Wine ½ drm.; Nitrate of Potash 10 grs.; Spirit of Nitrous Ether ½ drm.; Solution of Acetate of Ammonia 2 drms.; water to 1 oz. *City Chest.*

Haustus Salinus Aperiens.
Sulphate of Magnesia 2 drms.; Carbonate of Magnesia 20 grs.; Peppermint Water 1 oz. *Women.*

Haustus Salinus Effervescens.
Bicarbonate of Soda 20 grs.; water 1 oz.; Tartaric Acid 18 grs.; water 1 oz. (To be taken effervescing.) *Women.*

Haustus Scillæ.
Oxymel of Squill ½ drm.; Solution of Acetate of Ammonia 3 drms.; Spirit of Nitrous Ether 20 mins.; water to 1 oz. *Westminster.*

Haustus Scillæ Acidus.
Oxymel of Squill 1 drm.; Dil. Hydrochloric Acid 7 mins.; Spirit of Chloroform 10 mins.; Comp. Infusion of Gentian 1 oz. *Women.*

Haustus Scillæ Co.
Tincture of Squill 15 mins.; Carbonate of Ammonia 5 grs.; Comp. Tincture of Camphor 20 mins.; Spirit of Chloroform, 10 mins.; Infusion of Cascarilla to 1 oz. *Women.*

Tincture of Squill 15 mins.; Spirit of Nitrous Ether, ½ drm.; Solution of Acetate of Ammonia 2 drms.; Caraway Water to 1 oz. *Middlesex.*

Haustus Scoparii Co.
Spirit of Juniper 30 mins.; Tartrate of Potash 20 grs.; Decoction of Broom to 1 oz. *St. Bartholomew's.*

Acid Tartrate of Potash ½ drm.; Spirit of Nitrous Ether ½ drm.; Decoction of Broom 1 oz. *City Chest*

Acetate of Potash 20 grs.; Tincture of Squill 15 mins.; Spirit of Nitrous Ether ½ drm.; Decoction of Broom to 1 oz. *Middlesex.*

Haustus Senegæ et Ammoniæ.
Carbonate of Ammonia 5 grs.; Infusion of Senega 1 oz. *City Chest.*

Carbonate of Ammonia 5 grs.; Tincture of Squill 40 mins.; Tincture of Senega 40 mins.; water to 1 oz. *St. Bartholomew's.*

Haustus Senegæ Co.

Carbonate of Ammonia 4 grs.; Tincture of Squill 15 mins.; Pimento Oil 1 min.; Infusion of Senega to 1 oz. *Middlesex.*

Carbonate of Ammonia 5 grs.; Spirit of Chloroform 15 mins.; Tincture of Squill 10 mins.; Infusion of Senega to 1 oz. *Fever.*

Haustus Sennæ.

Sulphate of Magnesia 2 drms.; Compound Tincture of Senna (special form) 2 drms; Aromatic Spirit of Ammonia 20 mins.; Infusion of Senna to 1½ oz. *St. George's.*

Haustus Sennæ Co.

Alexandrian Senna Leaves 2 oz.; Ginger Root sliced 1½ drm.; Pimento contused 1½ drm.; boiling water 20 oz. Infuse 1 hour, strain, and add Sulphate of Magnesia 4 oz.; Caramel 1 fl. drm. Dose 1 to 2 oz. *Royal Free.*

Sulphate of Magnesia ¼ oz.; Oil of Peppermint ¼ min.; Infusion of Senna to 1½ oz. *St. Bartholomew's.*

Sulphate of Magnesia 120 grs.; Aromatic Spirit of Ammonia ½ drm.; Infusion of Senna 1½ oz. *City Chest.*

Sulphate of Magnesia 120 grs.; Tincture of Senna 2 drms.; Comp. Infusion of Senna to 1 oz. *Fever.*

Sulphate of Magnesia 120 grs.; Oil of Pimento 1 min.; Extract of Liquorice 10 grs.; Infusion of Senna to 1 oz. *Middlesex.*

Sulphate of Magnesia 160 grs.; Infusion of Senna to 1 oz. *London.*

Haustus Serpentariæ et Ammoniæ.

Carbonate of Ammonia 5 grs.; Infusion of Serpentary 1 oz. *City Chest.*

Carbonate of Ammonia 5 grs.; Spirit of Aniseed 10 mins.; Tincture of Serpentary 1 drm.; Spearmint Water to 1 oz. *St. Bartholomew's.*

Haustus Sodæ Citratis.
Bicarbonate of Soda 20 grs.; Citric Acid 17 grs.; water to 1 oz. Dissolve and mix. To be taken when effervescence has ceased. *St. Bartholomew's.*

Haustus Sodæ Hypophosphitis.
Hypophosphite of Soda 5 grs.; Spirit of Chloroform 10 mins.; Syrup ½ drm.; Camphor Water 1 oz. *City Chest.*

Haustus Sodæ Salicylatis.
Salicylate of Soda 15 grs.; water 1 oz. *Middlesex.*

Haustus Sodæ Tartratis.
Bicarbonate of Soda 20 grs.; Tartaric Acid 17 grs.; water 1 oz. Dissolve and mix. To be taken after effervescence has ceased. *St. Bartholomew's.*

Haustus Sodæ Tartratis Effervescens.
Bicarbonate of Soda 20 grs.; water to 1 oz. Dissolve and add Tartaric Acid 17 grs.; water to 1 oz. To be taken during effervescence. *St. Bartholomew's.*

Haustus Strychniæ Compositus.
Solution of Strychnia P. B. 4 mins.; Sulphate of Quinia 1 gr.; Sulphate of Zinc 1 gr.; Dil. Sulphuric Acid 2 mins.; water to 1 oz. *Middlesex.*

Haustus Strychniæ c. Ferro.
Solution of Strychnia 3 mins.; Solution of Perchloride of Iron 10 mins.; Syrup 30 mins.; water 1 oz. *Women.*

Solution of Strychnia 4 mins.; Tincture of Perchloride of Iron 10 mins.; water to 1 oz. *Middlesex.*

Solution of Strychnia 3 mins.; Tincture of Perchloride of Iron 15 mins.; Spirit of Chloroform 10 mins.; water 1 oz. *City Chest.*

Haustus Strychniæ c. Magnesiâ.
Sulphate of Iron 2 grs.; Sulphate of Magnesia 15 grs.; Solution of Strychnia 4 mins.; Spirit of Chloroform 10 mins.; Peppermint Water to 1 oz. *Middlesex.*

Haustus Taraxaci.

Juice of Dandelion 1 drm.; Sulphate of Magnesia ½ drm.; Peppermint Water to 1 oz. *Westminster.*

Haustus Taraxaci Acidus.

Juice of Dandelion ½ drm.; Dil. Nitro-hydrochloric Acid 15 mins.; Compound Infusion of Gentian to 1 oz. *Westminster.*

Haustus Terebinthinæ.

Oil of Turpentine 15 mins.; Mucilage 1 drm.; Pimento Draught to 1½ oz. *St. George's.*

Oil of Turpentine 15 mins.; Ether 15 mins.; Mucilage 2 drms.; Peppermint Water to 1 oz. *Fever.*

Oil of Turpentine 20 mins.; Mucilage 2 drms. Caraway Water to 1 oz. *St. Bartholomew's.*

Haustus Tragacanthæ c. Ipecacuanhâ.

Ipecacuanha Wine 10 mins.; Mucilage of Tragacanth 1 oz. *Fever.*

Haustus Valerianæ.

Ammoniated Tincture of Valerian 1 drm.; water to 1 oz. *Fever.*

Ammoniated Tincture of Valerian or Fœtid Spirit of Ammonia 20 mins.; Tincture of Hyoscyamus 30 mins.; Camphor Water 1 oz. *Samaritan.*

Infusion of Valerian ½ oz.; Camphor Water to 1 oz. *Westminster.*

Haustus Valerianæ Comp.

Carbonate of Ammonia 4 grs.; Infusion of Valerian 1 oz. *Women.*

Ammoniated Tincture of Valerian 20 mins.; Fœtid Spirit of Ammonia 15 mins.; Infusion of Quassia to 1 oz. *Middlesex.*

Haustus Valerianæ c. Hyoscyamo.

Ammoniated Tincture of Valerian ½ drm.; Tincture of Hyoscyamus ½ drm.; Camphor Water to 1 oz. *St. Bartholomew's.*

Haustus Zinci Emeticus.

Sulphate of Zinc 20 grs.; water to 1 oz. *St. Bartholomew's.*

Haustus Zinci Sulphatis.
>Sulphate of Zinc 20 grs. ; water 1 oz. *London.*

Haustus Zinci Sulphatis c. Ipecac.
>Sulphate of Zinc 20 grs. ; Ipecacuanha Wine ½ oz. ; water to 1½ oz. *London.*

HYDRARGYRI OLEAS.
Contains 10 per cent. of Oxide of Mercury. *Guy's.*

HYDRARGYRI OLEAS c. MORPHIA.
Morphia 1 gr. ; Oleate of Mercury 1 drm. *Guy's.*

INFUSA.

Infusum Gentianæ Aromaticum.
>(*Vel* Infusum Gentianæ Compositum.)
>Sliced Gentian ¼ oz. ; Lemon Peel 45 grs. ; Orange Peel 22½ grs.; boiling water 20 oz. *Consumption.*

Infusum Gentianæ Co.
>Sliced Gentian 60 grs. ; Powdered Canella 30 grs. ; boiling water 12 oz. : infuse 1 hour. Dose 1 to 2 oz. *St. George's.*
>Gentian Root 120 grs. ; Ginger 60 grs. ; boiling water 20 oz. : infuse 1 hour. Dose 1 to 2 oz. *London.*

Infusum Gentianæ Simplex.
>Gentian Root bruised 2 drms. ; boiling water 20 oz.: infuse 1 hour. Dose 1 to 2 oz. *Middlesex.*

Infusum Gummi Rubri.
>Red Gum ½ oz. ; boiling water 20 oz. *London.*

Infusum Tabaci.
>Leaf Tobacco 60 grs. ; boiling water 20 oz. *Women.*

INHALATIONS. (*See also* VAPORES.)

The following are for use with a Seigel's Steam Spray Producer:—

Inhalatio Acidi Carbolici.
>Carbolic Acid (No. 2 Calvert) 20 grs. ; hot water 1 oz. *Guy's.*

Inhalatio Acidi Sulphurosi.
> Sulphurous Acid 3 fl. drms.; water to 1 oz. *Guy's.*

Inhalatio Acidi Tannici.
> Tannic Acid 10 grs.; water 1 oz. *Guy's.*

Inhalatio Alkalina.
> Carbonate of Soda 10 grs.; water 1 oz. *Guy's.*

Inhalatio Aluminis.
> Alum 15 grs.; water 1 oz. *Guy's.*

Inhalatio Potassæ Chloratis.
> Saturated Solution of Chlorate of Potash. *Guy's.*

INJECTIONES.

Injectio Acidi Carbolici.
> Carbolic Acid 20 grs.; water 20 oz. *Women.*
> Carbolic Acid 5 grs.; water 1 oz. *Throat.*

Injectio Acidi Carbolici ad aures.
> Carbolic Acid (No. 2 Calvert) 12 grs.; hot water 1 oz. To be diluted with an equal quantity of warm water immediately before use. *London.*

Injectio Acidi Carbolici ad vaginam.
> Carbolic Acid (No. 2 Calvert) 2 grs.; hot water 2 drms. Dissolve and add water to 1 oz. *London.*

Injectio Acidi Tannici.
> Tannic Acid 2 drms.; Glycerine 1 oz.; water 3 oz. *Fever.*
> Tannic Acid 6 grs.; water 1 oz. *London.*

Injectio Alkalina.
> Carbonate of Potash 120 grs.; water 20 oz. *London Ophthalmic.*
> Solution of Potash 2 drms.; water 20 oz. *London. Women.*

Injectio Aluminis.
> Alum 60 grs.; water 20 oz. *Women.*
> Alum 5 grs.; water 1 oz. *Middlesex.*

Injectio Aluminis ad urethram.
> Alum 4 grs.; Carbolic Acid (No. 2 Calvert) $\frac{1}{4}$ gr.; water 1 oz. *London.*

Injectio Aluminis ad vaginam.
Alum 6 grs.; water 1 oz. *London.*

Injectio Aluminis Co.
Alum 3 grs.; Sulphate of Zinc 3 grs.; water 1 oz. *London.*

Injectio Aluminis et Acidi Tannici.
Alum 60 grs.; Tannic Acid 60 grs.; water 20 oz. *Women.*

Injectio Aluminis et Quercûs.
Alum 6 grs.; Decoction of Oak Bark 1 oz. *Royal Free.*

Injectio Aluminis et Zinci.
Alum 120 grs.; Sulphate of Zinc 40 grs.; water 20 oz. *British Skin.*
Alum 60 grs.; Sulphate of Zinc 60 grs.; water 20 oz. *Royal Free.*

Injectio Astringens.
Alum 4 grs.; Sulphate of Zinc 4 grs.; Tannic Acid 1½ gr.; water 1 oz. *London Ophthalmic.*

Injectio Boracis.
Borax 120 grs.; water 20 oz. *Women.*

Injectio Communis.
Alum 40 grs.; Sulphate of Zinc 24 grs.; Decoction of Oak Bark 4 oz. *Gt. Northern.*

Injectio Cupri Sulphatis.
Sulphate of Copper 1 gr.; water 1 oz. *London. Middlesex.*

Injectio Ferri Perchloridi.
Perchloride of Iron 60 grs.; water 1 oz. *Throat.*

Injectio Iodi.
Solution of Iodine 10 mins.; water 1 oz. *Throat.*

Injectio Plumbi.
Acetate of Lead 16 grs.; water 4 oz. *Fever.*
Acetate of Lead 30 grs.; water 20 oz. *Women.*

Diluted Solution of Subacetate of Lead 12 mins.; water to 1 oz. *London.*
Diluted Solution of Subacetate of Lead 1½ drm.; Rectified Spirit 2 drms.; water to 20 oz. *London Ophthalmic.*
Solution of Subacetate of Lead 1 drm.; water to 20 oz. *Middlesex.*
Solution of Subacetate of Lead 20 mins.; water to 1 oz. *Royal Free.*

Injectio Plumbi c. Opio.
Acetate of Lead 40 grs.; Extract of Opium 10 grs.; water 20 oz. *Women.*

Injectio Potassæ Permanganatis.
Solution of Permanganate of Potash 6 mins.; water to 1 oz. *Throat. Women.*
Solution of Permanganate of Potash 24 mins.; water to 1 oz. *London.*
Permanganate of Potash 10 grs.; Distilled Water 20 oz. *Charing Cross.*

Injectio Quercûs Co.
Alum 4 grs.; Sulphate of Zinc 4 grs.; Decoction of Oak Bark 1 oz. *London.*
Alum 80 grs.; Sulphate of Zinc 80 grs.; Decoction of Oak Bark 10 oz.; water 10 oz. *Middlesex.*

Injectio Quercûs c. Alumine.
Alum 1 drm.; Decoction of Oak Bark 4 oz. *Fever.*

Injectio Quiniæ Sulphatis.
Sulphate of Quinia 4 grs.; Diluted Sulphuric Acid 3 mins.; water 1 oz.; to be diluted with 1 oz. of warm water before using for the ear. *University.*

Injectio Thymol.
Thymol 1 gr.; Rectified Spirit 1 drm.; Glycerine 1 drm.; water to 1 oz. *London.*

Injectio Zinci Chloridi.
Chloride of Zinc 2 grs.; water 1 oz. *Royal Free.*
Chloride of Zinc 1 gr.; water 1 oz. *London. Middlesex.*

Injectio Zinci Sulphatis ad urethram.
Sulphate of Zinc 3 grs.; water 1 oz. *London.*

Injectio Zinci Sulphatis.
: Sulphate of Zinc 2 grs.; water to 1 oz. *Charing Cross. Middlesex.*
Sulphate of Zinc 2 grs.; Alum 4 grs.; water 1 oz. *King's.*
Sulphate of Zinc 3 grs.; water 1 oz. *Royal Free. Women.*
Sulphate of Zinc 6 grs.; water 1 oz. *London.*
Sulphate of Zinc 8 grs.; water 1 oz. *Fever.*

Injectio Zinci Sulphatis et Acidi Carbolici.
: Sulphate of Zinc 4 grs.; Carbolic Acid 4 grs.; water 1 oz.; to be diluted with 1 oz. of warm water before using for the ear. *University.*

Injectio Zinci Sulphocarbolatis.
: Sulphocarbolate of Zinc 2 grs.; water 1 oz. *King's.*
Sulphocarbolate of Zinc 3 grs.; water 1 oz. *Charing Cross.*
Sulphocarbolate of Zinc 6 grs.; water 1 oz. *London.*

INJECTIONES HYPODERMICÆ.

Injectio Acidi Acetici Hypod.
: Diluted Acetic Acid 15 to 30 mins. *Throat.*

Injectio Acidi Carbolici Hypod.
: Carbolic Acid (No. 1 Calvert) 15 grs.; Rectified Spirit 15 mins.; water to 1 drm. Dose 1 to 4 mins. $= \frac{1}{4}$ to 1 gr. acid. *London.*

Injectio Acidi Sclerotici Hypod.
: Sclerotic Acid pure 15 grs.; Glycerine pure 15 mins.; water to 1 drm. Dissolve and add 1 per cent. of Carbolic Acid No. 1. Dose 1 to 4 mins. $= \frac{1}{4}$ to 1 gr. acid. *London.*

The following remedies in the form of Hypodermic Discs are used (*London*) :—

Discs of Aconitia $\frac{1}{60}$ gr. in each.
,, Apomorphia $\frac{1}{10}$ gr. in each.
,, Atropia (Sulphate) $\frac{1}{120}$ gr. in each.
,, Caffeine $\frac{1}{2}$ gr. in each.
,, Codeia $\frac{1}{4}$ gr. in each.
,, Curara $\frac{1}{10}$ gr. in each.
,, ,, $\frac{1}{20}$ gr. in each.

Discs of Daturine $\frac{1}{120}$ gr. in each.
"   Digitaline.
"   Ergotine.
"   Gelsemium, equal to 10 mins. Tincture.
"   Hyoscyamine $\frac{1}{300}$ gr. in each.
"   Morphia $\frac{1}{6}$ gr. in each.
"   "   and Atropine, $\frac{1}{6}$ gr. Morphia and $\frac{1}{120}$ gr. Atropine in each.
"   Physostigmine (or Eserine) equal to $\frac{1}{8}$ gr. Extract of Calabar Bean.
"   Pilocarpine $\frac{1}{12}$ gr. in each.
"   Sclerotic Acid $\frac{1}{10}$ gr. in each.
"   Strychnia $\frac{1}{60}$ gr. in each.

Injectio Atropiæ Hypod.

Sulphate of Atropia $\frac{1}{2}$ gr.; Glycerine pure 15 mins.; water to 1 drm. Dissolve and add 1 per cent. of Carbolic Acid No. 1. Dose 1 to 2 mins. = $\frac{1}{120}$ to $\frac{1}{60}$ Sulphate. *London.*

Sulphate of Atropia 1 gr.; Distilled Water 1 oz. (4 mins. = $\frac{1}{120}$ gr. Atropia). *Women.*

Injectio Atropiæ Sulphatis.

Sulphate of Atropia $\frac{1}{4}$ gr.; water 1 drm. Dose 2 to 4 mins. *London Ophthalmic.*

Injectio Curare Hypod.

Curare 4 grs.; distilled water 1 fl. dr. Rub the Curare with the water, allow it to stand for 48 hours, and then filter. *Guy's.*

Curare 5 grs.; Glycerine pure 15 mins.; water to 1 drm. Mix the glycerine and water and heat them to the boiling-point; rub the Curare with the hot liquid in a mortar, triturate frequently during 24 hours and then filter. Dose 1 min. = $\frac{1}{12}$ gr. Curare, cautiously increased. *London.*

Injectio Ergotinæ Hypod.

Ergotine $\frac{1}{6}$ gr.; Glycerine 3 mins.; water 3 mins. *Westminster.*

Ergotina (Bonjean's) 10 grs.; Distilled Water 1 fl. dr.; dissolve. *Guy's.*

Ergotine 240 grs.; Distilled Water sufficient to make up 1 fl. oz.; in this dissolve Carbolic Acid 4 grs.; 2 to 6 mins. *Throat.*

Ergotine 15 grs.; Glycerine pure 15 mins.; water to 1 drm. Dissolve and add 1 per cent. Carbolic Acid No. 1. Dose 1 to 4 mins. = $\frac{1}{4}$ to 1 gr. Ergotine. To be injected deeply. *London.*

Injectio Ergotæ Hypod.
Bonjean's Liquoris Ergotæ (6 mins. = 1 gr). *Women.*

Injectio Hydrargyri Iodidi Hypod.
Red Iodide of Mercury 7½ grs.; Iodide of Sodium about 7½ grs., or less if possible; Distilled water to 1 oz. Dissolve the Iodide of Sodium in a little water and add it drop by drop to the Iodide of Mercury till the latter is all dissolved. Dilute with more water and filter, pouring over the filter a sufficient quantity of water to produce 1 oz. of fluid. This will contain $\frac{1}{16}$ gr. of Iodide of Mercury in 4 mins. Dose 4 to 10 mins. *Throat.*

Injectio Hydrarg. Perchloridi Hypod.
Corrosive Sublimate ½ gr.; Chloride of Sodium pure 5 grs.; water to 1 drm. Dose 4 to 12 mins. = $\frac{1}{30}$ to $\frac{1}{10}$ gr. in divided portions in the course of one day. *London.*

Injectio Iodi Hypod. Communis.
Tincture of Iodine B. P. 15 to 30 mins. *Throat.*

Injectio Iodi Hypod. Fortior.
Iodine 40 grs.; Absolute Alcohol 1 oz. 10 to 15 mins. *Throat.*

Injectio Iodi Hypod. Fortissima.
Iodine 360 grs.; Iodide of Potassium 360 grs.; Distilled Water to 1 oz.; dissolve. 3 to 5 mins. *Throat.*

Injectio Morphiæ Hypod.
Acetate of Morphia 80 grs.; Distilled Water 1 oz.; dissolve. 6 mins. contain 1 gr. *Throat.*

Acetate of Morphia 40 grs.; Distilled Water 1 oz. Brit. Pharm. 12 mins. contain 1 gr. *London Ophthalmic. Middlesex. Women.*

Acetate of Morphia 10 grs.; Glycerine 15 mins.; Acetic Acid, Solution of Potash, and Distilled Water of each *q. s.* Dissolve the Morphia in ½ drm. of the water with the aid of a gentle heat and a few drops of Acetic Acid; when cool add Solution of Potash nearly to neutralization, then mix with the Glycerine and sufficient water to make 1 drm.; lastly add 1 per cent. Carbolic Acid No. 1. Dose 1 to 3 mins. = ⅙ to ½ gr. Acetate of Morphia. *London.*

Injectio Morphiæ Hypod. Fortior.
Acetate of Morphia 80 grs.; Distilled Water 1 oz. *Middlesex.*

Injectio Morphiæ et Atropiæ Hypod.
Injectio Morphiæ Hypoderm. London Hospital 2 drms.; Injectio Atropiæ Hypod. London Hospital 1 drm.; mix. Dose 3 mins. = ⅓ gr. Acetate of Morphia, $\frac{1}{120}$ gr. Sulphate of Atropia. *London.*

Sulphate of Atropia ⅛ gr.; Hypodermic Injection of Morphia P. B. 1 drm. Dose 1 to 4 mins. *London Ophthalmic. Middlesex.*

Injectio Morphiæ Hypoderm. 1 drm.; Solution of Atropia 20 mins. (4 mins. = Morphia ¼ gr. Atropia $\frac{1}{120}$ gr.) *Women.*

Injectio Quiniæ Hypod.
Neutral Sulphate of Quinine 20 grs.; Glycerine pure 1 drm.; dissolve by the aid of heat. Dose 1 to 4 mins. = ¼ to 1 gr. Sulphate of Quinine. This solution should be warmed before use. *London.*

Injectio Strychniæ Hypod.
Solution of Strychnia Brit. Pharm. Dose 1 to 2 mins. = $\frac{1}{120}$ to $\frac{1}{60}$ gr. Strychnia. *London.*

INSUFFLATIONES.

Insufflatio Acidi Tannici.
Tannic Acid in powder 2 grs.; Starch in powder ½ gr. An astringent in hæmorrhage from the larynx and trachea. *Throat.*

Insufflatio Acidi Tannici et Iodoformi. (*Nasal.*)
Tannic Acid in powder 5 grs.; Iodoform in fine powder 2 grs.; Gum Acacia in fine powder 3 grs. An astringent and alterative in post-nasal catarrh. *Throat.*

Insufflatio Aluminis.
Alum in fine powder 3 grs.; Starch in powder ½ gr. A mild astringent in chronic tracheitis. *Throat.*

Insufflatio Aluminis. (*Aural.*)
Alum in fine powder, Subnitrate of Bismuth in fine powder. Of each an equal quantity. Astringent. *Throat.*

Insufflatio Ammonii Chloridi.
Chloride of Ammonium in fine powder 2 grs.; Starch in powder ½ gr.; mix. A solvent when mucus is adherent. *Throat.*

Insufflatio Argenti Nitratis. (*Aural.*)
Nitrate of Silver in fine powder 6 grs.; Subnitrate of Bismuth in fine powder 60 grs. Caustic. *Throat.*

Insufflatio Bismuthi.
Subnitrate of Bismuth 2 grs.; Starch in powder ½ gr. A sedative in congestion of the trachea. *Throat.*

Insufflatio Bismuthi, Morphiæ, et Iodoformi. (*Nasal.*)
Carbonate of Bismuth 7 grs.; Acetate of Morphia ¼ gr.; Iodoform in fine powder 5 grs.; Gum Acacia in fine powder 5 grs. *Throat.*

Insufflatio Boracis.
Borax in powder 3 grs.; Starch in powder ½ gr. Mildly detergent. *Throat.*

Insufflatio Iodoformi.
Iodoform in fine powder 1 gr.; Starch in powder ½ gr. Antiseptic and mildly caustic. *Throat.*

Insufflatio Iodoformi. (*Aural.*)
Iodoform in fine powder; Subnitrate of Bismuth in fine powder. Of each an equal quantity. Antiseptic and mildly caustic. *Throat.*

Insufflatio Morphiæ.
Acetate of Morphia $\frac{1}{16}$ gr.; Starch in powder ½ gr. Rub together to form a powder. Insufflations of Morphia are also used containing respec-

tively ⅛, ¼, and ½ gr. of Acetate of Morphia in
each, combined with ½ gr. of Starch. Sedative,
especially in laryngeal phthisis. *Throat.*

Insufflatio Plumbi Acetatis.
Acetate of Lead in fine powder 1 gr.; Starch in
powder ½ gr. Styptic. *Throat.*

LAPIS DIVINUS.
Sulphate of Copper 3 oz.; Nitrate of Potash 3 oz.;
Alum 3 oz.; Camphor 60 grs. *Guy's.*

LINCTI.

Linctus Acidi Tanno-Gallici.
Tannic Acid 6 drms.; Gallic Acid 2 drms.; water 1
oz. *St. Bartholomew's.*

Linctus Acidus.
Dil. Sulphuric Acid 5 mins.; Oxymel 20 mins.:
Spirit of Chloroform 2 mins.; Treacle to 1 drm.
University.
Dil. Sulphuric Acid 5 mins.; Oxymel 25 mins.;
Simple Linctus to 1 drm. *Consumption.*

Linctus Acidus c. Opio.
Tincture of Opium 15 mins.; Diluted Sulphuric
Acid 15 mins.; Glycerine 1 oz.; Camphor
Water 1 oz. Dose 1 to 2 drms. *Gt. Northern.*

Linctus Ammoniæ.
Tincture of Squill 5 mins.; Carbonate of Ammonia
⅛ gr.; Essence of Anise P. B. 1 min.; Borax
¼ gr.; Mucilage 30 mins.; water to 1 drm.
Consumption.

Linctus Communis.
Acetic Acid 2 mins.; Vinegar of Squill 36 mins.;
Syrup of Poppies 36 mins.; Confection of Hips
100 grs.; Powdered Tragacanth 6 grs.; boiling
water to 1 oz. Dose 1 teaspoonful. *St. Bartholomew's.*
Comp. Tincture of Camphor 15 mins.; Oxymel of
Squill 45 mins. Dose 1 drm. *Women.*
Dil. Sulphuric Acid 7½ mins.; Tincture of Squill
30 mins.; Syrup of Extract of Poppies 1½
drm.; Powdered Tragacanth 6 grs.; Treacle

2 drms.; water to 1 oz. Dose ½ to 2 drms. *Middlesex.*

Dil. Sulphuric Acid 15 mins.; Syrup of Squill 30 mins.; Paregoric 2 drms.; Treacle 1 drm.; Anise Water to 1 oz.; mix. Dose 1 drm. *King's.*

Dil. Sulphuric Acid 30 mins.; Vinegar of Squill 3 drms.; Tincture of Opium 30 mins.; Powdered Tragacanth 40 grs.; Treacle 2 oz.; water 2 oz. *London Ophthalmic.*

Dil. Sulphuric Acid 40 mins.; Tincture of Opium 20 mins.; Treacle 300 grs.; Hot water to 1 oz. Dose 1 drm. *Throat.*

Dil. Sulphuric Acid 5 drms.; Syrup of Squill 5 drms.; Syrup of Poppies 5 drms.; Comp. Tincture of Camphor 5 drms.; Ipecacuanha Wine 2 drms.; Gum Arabic 5 drms.; Treacle 2½ fl. oz.; water 7½ oz. Dose 1 to 2 drms. *Westminster Ophthalmic.*

Olive Oil 4 drms.; Confection of Hips 6 drms.; Vinegar of Squill 1½ drm.; Tincture of Opium 7½ mins.; Treacle 3 drms. Dose 1 to 2 drms. *St. George's.*

Solution of Hydrochlorate of Morphia 3 mins.; Spirit of Chloroform 3 mins.; Glycerine and water of each ½ drm. *City Chest.*

Linctus Creasoti.

Creasote 1 min.; Glycerine 56 mins.; water to 1 oz. *Royal Chest.*

Linctus Glycerinæ.

Compound Tincture of Camphor 2 drms.; Vinegar of Ipecacuanha 40 mins.; Glycerine ½ oz.; water to 1 oz. Dose 1 drm. *Throat.*

Linctus c. Ipecacuanhâ.

Ipecacuanha Wine 30 mins.; Linctus to 1 oz. Dose 1 teaspoonful. *St. Bartholomew's.*

Linctus Ipecacuanhæ c. Morphiâ.

Solution of Hydrochlorate of Morphia 2 mins.; Spirit of Chloroform 2 mins.; Ipecacuanha Wine 2 mins.; Mucilage of Acacia 30 mins.; water to 1 drm. *Charing Cross.*

Linctus Limonis.

Syrup of Lemon 6 drms.; Solut. Acet. Morphia 2 fl. drms.; water to 3 oz. Dose 1 drm. *Throat.*

Linctus Morphiæ.

Solution of Hydrochlorate of Morphia 3 mins.; Spirit of Chloroform 3 mins.; Glycerine 1 drm.; *Westminster.*

Solution of Hydrochlorate of Morphia 3 mins.; Spirit of Chloroform 3 mins.; Treacle, Honey, or Glycerine 60 grs.; water to 1 drm. *University.*

Solution of Hydrochlorate of Morphia 5 mins.; Diluted Hydrocyanic Acid $2\frac{1}{2}$ mins.; Spirit of Ether $2\frac{1}{2}$ mins.; Syrup $\frac{1}{2}$ drm.; water to 1 drm. Dose $\frac{1}{2}$ to 1 drm. *Royal Free.*

Hydrochlorate of Morphia $\frac{1}{8}$ gr.; Dil. Hydrochloric Acid 2 mins.; Syrup of Squill 20 mins.; Dil. Hydrocyanic Acid 2 mins.; water to 1 drm. *Royal Chest.*

Muriate of Morphia 1 gr.; Dil. Hydrochloric Acid 2 mins.; Dil. Prussic Acid 30 mins.; Syrup of Squill 2 oz.; water to 4 oz.; Dose $\frac{1}{2}$ to 2 drms. *Middlesex.*

Linctus Morphiæ Acetatis.

Solution of Acetate of Morphia 8 mins.; Chloric Ether 3 mins.; Lemon Juice 15 mins.; Mucilage to 1 drm. *Consumption.*

Acetate of Morphia $\frac{1}{12}$ gr.; Oxymel 1 drm.; water 1 drm. Dose 2 drms. *St. Mary's.*

Linctus Morphiæ c. Chloroformo.

Solution of Hydrochlorate of Morphia 24 mins.: Chloroform 8 mins.; Rectified Spirit 72 mins.; Glycerine to 1 oz. Dose 1 teaspoonful. *St. Bartholomew's.*

Linctus Morphiæ Compositus.

Hydrochlorate of Morphia $\frac{1}{16}$ gr.; Laurel Water 8 mins.; Spirit of Chloroform 4 mins.; Syrup of Lemons 30 mins.; Glycerine to 1 drm. Dose 1 drm. *London.*

Linctus Opiatus.

Tincture of Opium 10 mins.; Dil. Sulphuric Acid 10 mins.; Treacle 6 drms.; Water to 1 fl oz. Dose 1 to 2 teaspoonfuls. *Guy's.*

Tincture of Opium 16 mins.; Dil. Sulphuric Acid 32 mins.; Treacle 160 mins.; boiling water to 1 oz.; Dose 1 to 2 drms. *London.*

Tincture of Opium 15 mins.; Dil. Sulphuric Acid 10 mins.; Treacle 4 fl. drms.; Water to 1 oz. Dose 1 drm. *Royal Free.*

Linctus Opii.
Tincture of Opium 5 mins.; Dil. Sulphuric Acid 5 mins.; Simple Linctus to 1 drm. *Consumption.*

Linctus Papaveris.
Syrup of Poppies 1 oz.; Syrup of Tolu 1 oz.; Compound Tincture of Camphor 1 oz. Dose 1 drm. *Throat.*

Linctus Scillæ.
Oxymel of Squill 20 mins.; Dil. Sulphuric Acid 5 mins.; Tincture of Opium 2 mins.; Simple Linctus to 1 drm. *Consumption.*
Oxymel of Squill 1 drm.; Paregoric 15 mins. Mucilage 1 drm. Dose 2 drms. *St. Mary's.*
Syrup of Squill, Syrup of Poppies, Syrup of Lemons, Syrup of Tolu, equal quantities. Dose 1 drm. *Throat.*

Linctus Scillæ c. Opio.
Oxymel of Squill 20 mins.; Liquid Extract of Opium 2 mins.; Dil. Sulphuric Acid 5 mins.; Treacle 20 mins; water to 1 drm. *Charing Cross.*

Linctus Scillæ Co. vel c. Opio.
Oxymel of Squill 5 mins.; Compound Tincture of Camphor 2½ mins.; Spirit of Nitrous Ether 2½ mins.; water to 1 drm. *Westminster.*
Oxymel of Squill 20 mins.; Comp. Tincture of Camphor 10 mins.; Ipecacuanha Wine 5 mins.; Mucilage to 1 drm. *Royal Chest.*
Oxymel of Squill 24 mins.; Comp. Tincture of Camphor 12 mins.; Ipecacuanha Wine 6 mins.; Mucilage to 1 drm. *University.*

Linctus Simplex vel Theriaca Preparata.
Treacle 20 mins.; Spirit of Chloroform 2 mins.; water to 1 drm. *Consumption.*

Linctus pro Tussi.
Oxymel of Squill, Syrup of Poppies, Mucilage, of each equal parts. *St. Thomas's.*

LINIMENTA.

Linimentum Acidi Sulphurici.
Strong Sulphuric Acid ½ fl. drm.; Spirit of Turpentine 4 drms.; Olive Oil 12 drms.; mix the Oils and add the Acid gradually. *St. George's.*

Linimentum Aconiti.
Tincture of Aconite 2 drms.; Glycerine 4 drms.; water ½ oz.; *Samaritan.*

Linimentum Aconiti Compositum.
Tincture of Aconite 5 drms.; Belladonna Liniment 2 drms.; Chloroform 1 drm. *Westminster Ophthalmic.*

Linimentum Aconiti Mitius.
Aconite Liniment B.P. 1 oz.; Soap Liniment 3 oz. *London.*

Linimentum Aconiti et Opii.
Tincture of Aconite 3 drms.; Liniment of Opium to 1 oz. *St. Thomas's.*

Linimentum Aconiti et Saponis.
Liniment of Aconite 1 oz.; Soap Liniment 3 oz. *London Ophthalmic.*

Linimentum Æruginis.
Powder of Verdigris 1 oz.; Dil. Acetic Acid 7 oz. Honey 14 oz.; dissolve the Verdigris in the Acid, and filter; add the Honey, and boil to a proper consistence. *Middlesex.*

Linimentum Ammoniæ.
Solution of Ammonia 1 oz.; Olive Oil 3 oz. *London.*

Linimentum Ammoniæ Mitius.
Solution of Ammonia 1 drm.; Olive Oil 7 drms. *City Chest.*

Linimentum Belladonnæ et Chloroformi.
Liniment of Belladonna 6 drms.; Chloroform 2 drms.; Compound Camphor Liniment 1 oz. *Westminster. Westminster Ophthalmic.*

Linimentum Belladonnæ Compos.
: Extract of Belladonna 1 drm.; Distilled Water 1 drm.; Soap Liniment 6 drms. *London Ophthalmic.*

Linimentum Belladonnæ c. Glycerino.
: Extract of Belladonna 1 oz.; water ½ oz.; Glycerine 3½ oz. *Gt. Northern.*

Linimentum Belladonnæ Mitius.
: Belladonna Liniment B.P. 1 oz.; Soap Liniment 3 oz. *London.*

Linimentum Calcis.
: Solution of Lime B.P. and Linseed Oil equal parts. *London. Middlesex.*

Linimentum Calcis Co.
: Olive Oil 1 oz.; Oil of Turpentine 1 drm.; Lime Water to 3 oz. *Gt. Northern.*
: Oil of Turpentine 3 drms.; Olive Oil 4 oz.; Solution of Lime to 12 oz. *Royal Free.*

Linimentum Camphoræ c. Cantharide.
: Liniment of Camphor 14½ drms.; Tincture of Opium 2 drms.; Tincture of Cantharides 3½ drms. *Women.*

Linimentum Cantharidis.
: Tincture of Cantharides 2 drms.; Liniment of Soap 6 drms. *City Chest.*
: Vinegar of Cantharides 6 drms.; Glacial Acetic Acid 40 mins.; Glycerine 6 drms.; Spirit of Rosemary 20 mins.; water to 4 oz. *St. Bartholomew's.*

Linimentum Cantharidis c. Ammoniâ.
: Almond Oil 6 drms.; Strong Solution of Ammonia 160 mins.; Glycerine 160 mins.; Tincture of Cantharides 80 mins.; Spirit of Rosemary 20 mins.; water to 4 oz. *St. Bartholomew's.*

Linimentum Cantharidis c. Opio.
: Tincture of Cantharides 2 drms.; Tincture of Opium 2 drms.; Strong Solution of Ammonia 4 drms.; Liniment of Soap 1 oz. *London.*

Linimentum Cantharidis et Camphoræ.
: Vinegar of Cantharides, Acetic Acid, Spirit of Camphor, of each equal parts. *Royal Chest.*

Linimentum Chloroformi.
Chloroform 1 oz.; Olive Oil 2 oz.; mix. *London.*

Linimentum Commune.
Soft Soap 3; Camphor 1; Rectified Spirit 8; water 8. *St. George's.*
Olive Oil 1 oz.; Solution of Ammonia ½ oz. *Samaritan.*

Linimentum Cretæ.
Prepared Chalk 2 oz.; Olive Oil 6 drms.; Rose Water 6 drms. *St. George's.*

Linimentum Crotonis. (*See also* Lin. Olei Crotonis.)
Croton Oil 1 drm.; Soap Liniment to 2 oz. *Middlesex. Royal Chest.*
Croton Oil 1 drm; Olive Oil to 1 oz. *St. Mary's.*
Croton Oil 1 drm.; Oil of Turpentine 3 drms.; Olive Oil 4 drms. *London Ophthalmic.*

Linimentum Crotonis Dilutum.
Liniment of Croton Oil P. B., Liniment of Soap, equal parts. *Consumption.*
Croton Oil 1 drm.; Soap Liniment to 2 oz. *University.*

Linimentum Hydrarg. Oleatis (10 per cent.).
Precipitated Peroxide of Mercury 1 drm.; Oleic Acid 10 drms.: keep agitating the Acid in the mortar; sprinkle in the Peroxide gradually, and triturate the mixture frequently during 24 hours, until the Peroxide is dissolved, and a viscid solution is formed. *London Ophthalmic. Throat. University.*

Linimentum Hydrarg. Oleatis (5 per cent., 10 per cent., and 15 per cent.).
Yellow Oxide of Mercury, freshly prepared, 24 grs., 48 grs., and 72 grs., to 1 oz. of Oleic Acid. Dissolve with a gentle heat. *St. Bartholomew's.*

Linimentum Hydrarg. Oleatis (5 per cent.).
Yellow Oxide of Mercury, recently precipitated, 1 drm.; Oleic Acid 5 drms.; Olive Oil to 20 drms. Sprinkle in the Oxide to the Acid in a porcelain dish, raise the temperature to 120° F. and keep it at this, constantly stirring, till they combine; then add the Oil, and mix well together. *London.*

Linimentum Hydrarg. Oleatis c. Morphiâ.
> Liniment of Oleate of Mercury 10 per cent.; Liniment of Oleate of Morphia 2 per cent.: of each equal parts. *London.*
> Pure Morphia 10 grs.; Oleic Acid 5 drms.: dissolve, and add Liniment of Oleate of Mercury 5 drms. *London Ophthalmic. Throat. University.*

Linimentum Iodi Co.
> Tincture of Iodine $2\frac{1}{2}$; Strong Solution of Ammonia 1; Common Liniment $2\frac{1}{2}$. *St. George's.*

Linimentum Iodi c. Belladonnâ.
> Iodine Liniment, Belladonna Liniment, equal parts. *Women.*

Linimentum Morphià Oleatis (2 per cent.).
> Morphia (Alkaloid) 12 grs.; Oleic Acid 10 drms: dissolve by aid of heat. *London.*

Linimentum Myristicæ.
> Volatile Oil of Nutmeg 1 oz.; Olive Oil 3 oz. *London.*
> Expressed Oil of Nutmeg 1 oz.; Olive Oil 3 fl. oz. *Guy's.*

Linimentum Olei Crotonis.
> Croton Oil 1 drm.; Soap Liniment to $1\frac{1}{2}$ oz. *Gt. Northern.*

Linimentum Olei Crotonis et Camphoræ.
> Oil of Croton 1 drm.; Liniment of Camphor 15 drms. *City Chest.*

Linimentum Opii.
> Tincture of Opium 1 oz.; Liniment of Soap 3 oz. *London.*
> Tincture of Opium 1 oz.; Ammoniated Liniment of Soap to 4 oz. *St. Bartholomew's.*

Linimentum Opii et Chloroformi.
> Tincture of Opium 2 drms.; Chloroform 2 drms.; Soap Liniment to 1 oz. *Gt. Northern.*
> Tincture of Opium 2 drms.; Chloroform 2 drms.; Compound Camphor Liniment 4 drms.; Liniment of Soap 1 oz. *City Chest.*

Linimentum Plumbi c. Oleo.
> Solution of Subacetate of Lead $\frac{1}{2}$ oz.; Olive Oil to 4 oz. *St. Bartholomew's.*

Linimentum Potassii Iodidi.
Iodide of Potassium 1 drm; Stronger Solution of Ammonia 2 drms.; Liniment of Soap to 2 oz. *City Chest.*

Linimentum Saponis.
Soft Soap 1 oz.; hot water 10 oz. *Middlesex.*
Soft Soap 2½ oz.; Camphor 1 oz.; Rectified Spirit 2 oz.; boiling water to 20 oz. *Gt. Northern.*
Soft Soap 6 oz.; Camphor 1½ oz.; Rectified Spirit 6 oz.; Oil of Turpentine 1 oz.; water 30 oz. *London.*

Linimentum Saponis Comp.
Soft Soap 4 lbs.; Camphor 1 lb.; Methylated Spirit of Wine 3 galls.; water to 5 galls. *Royal Free.*

Linimentum Saponis c. Ammoniâ.
Soft Soap ½ oz.; Solution of Ammonia 1 drm.; Methylated Spirit 3 drms.; Oil of Marjoram 6 mins.; Oil of Turpentine 1½ drm.; boiling water to 4 oz. *St. Bartholomew's.*

Linimentum Saponis et Iodi.
Tincture of Iodine 1 drm.; Liniment of Soap 7 drms. *City Chest.*

Linimentum Saponis Mollis.
Soft Soap 12 drms.; Camphor 4 drms.; Oil of Origanum 1 drm.; Rectified Spirit 7½ oz. *Guy's.*

Linimentum Saponis c. Opio.
Tincture of Opium 2 drms.; Soap Liniment 10 drms. *Consumption.*
Tincture of Opium 2 drms.; Soap Liniment 6 drms. *Middlesex.*

Linimentum Sinapis.
Mustard 24 grs.; Oil of Turpentine 1 oz. *St. Thomas's.*

Linimentum Terebinth.
Oil of Turpentine 1½ oz.; Soap Liniment 1½ oz. *London.*

Linimentum Terebinth. c. Ammoniâ.
Liniment of Turpentine 6 drms.; Solution of Ammonia 6 drms.; Oil of Cajeput 15 mins.; Olive Oil to 2 oz. *Royal Chest.*
Liniment of Turpentine 6 drms.; Solution of Ammonia 6 drms.; Oil of Cajeput 15 mins.; Olive Oil 2 oz. *Middlesex.*

Linimentum Terebinth. c. Chloroformo.
Liniment of Turpentine 1½ oz.; Chloroform ½ oz. *St. Mary's.*

Linimentum Tiglii.
Croton Oil and Turpentine of each equal parts *Samaritan.*

LINTEA MEDICATA.

Linteum Boracicum.
Lint steeped in a hot saturated watery solution of Boracic Acid, and dried without wringing. *Guy's. London.*

Linteum Carbolicum.
Lint steeped in a hot solution of Carbolic Acid, 1 part of acid, No. 2 Calvert, to 49 parts of hot water, and dried without wringing. *London.*

LIQUORES.

Liquor Argenti Nitratis.
Nitrate of Silver 5, 10, 15, or 20 grs.; water 1 oz. *London Ophthalmic.*

Liquor Argenti Nitratis Opiatus.
Nitrate of Silver 2 grs.; Wine of Opium 2 oz. *St. George's.*

Liquor Arsenici et Hydrargyri Hydriodatis.
(Donovan's Solution.)
Each fluid drm. contains $\frac{1}{12}$ gr. Arsenic; ¼ gr. Mercury; ¾ gr. Iodine. Dose 10 to 30 mins. *London.*

Liquor Atropiæ Salicylatis.
Atropine 5 grs.; Salicylic Acid 7½ grs.; water 10 oz. *Charing Cross.*

Liquor Belladonnæ.
　Extract of Belladonna ½ oz.; Proof Spirit 1 drm.;
　water 15 drms. *St. George's.*

Liquor Bituminis Comp.
　Coal Tar 1 oz.; Boiling Water 2 oz.; shake well
　and add Tincture of Quillaia 20 oz. Agitate
　occasionally in a closed vessel for 12 hours and
　filter. *London.*

Liquor Chlori.
　Chlorate of Potash 30 grs.; Hydrochloric Acid ½
　oz.; water to 1 oz. *London.*

Liquor Epispasticus.
　Cantharides in powder 8 oz.; Glacial Acetic Acid
　2 oz.; Ether *q. s.* to percolate 20 oz. *London
　Ophthalmic.*

Liquor Ferri Acetatis.
　Same strength as Tinct. Ferri Acet P. B. *London.*

Liquor Ferri et Ammoniæ Citratis. (Vin. Ferri.)
　Citrate of Iron and Ammonia 8 grs.; Treacle ½
　drm.; water to 1 oz. Dose 1 to 4 teaspoonfuls.
　St. Bartholomew's.

Liquor Ferri Dialysatus.
　Each drm. contains 3 grs. of Oxide of Iron. Dose
　10 to 20 mins. *London.*

Liquor Hydrargyri Bibromidi.
　Bibromide of Mercury 1 gr.; Distilled Water 2 oz.
　(Dose same as Liq. Hydrarg. Perchlor.) *Women.*

Liquor Iodoformi Æthereus.
　A saturated solution of Iodoform in Ether. *Guy's.*

Liquor Picis Alkalinus.
　Caustic Potash 60 grs.; Distilled Water 5 drms.;
　mix and add gradually Stockholm Tar 2 drms.
　Charing Cross.

Liquor Ruber.
　Raw Sugar 1 oz.; boiling water 20 oz. Heat the
　Sugar in an iron vessel until it fuses to a deep
　brown colour, then add the boiling water.
　London.

LOHOCH. See LINCTI.
LOTIONES.

Lotio Aceti.
Vinegar 5 drms.; water to 10 oz. *St. Bartholomew's.*
Dil. Acetic Acid 5 drms.; Distilled Water 10 oz. *London Ophthalmic. Westminster Ophthalmic.*

Lotio Acidi Acetici.
Dil. Acetic Acid 2 oz.; water to 20 oz. *City Chest.*

Lotio Acidi Benzoici.
Benzoic Acid 2 grs.; Rectified Spirit 24 mins.; water to 1 oz. *Middlesex.*
Benzoic Acid 1 gr.; water 1 oz. *British Skin.*

Lotio Acidi Boracici.
A saturated Aqueous Solution. *London. University.*
Boracic Acid 15 grs.; water 1 oz. *British Skin.*
Boracic Acid 1 oz.; Cochineal *q. s.*; water to 40 oz. *Royal Free.*

Lotio Acidi Carbolici.
Carbolic Acid 1 part; water at 122° F., to 20, 40, 80, or 100 parts as required. *University.*
Carbolic Acid No. 2 Calvert 1 part; Boiling Water to 20, 40, 80, or 100 parts as required. *London.*
Carbolic Acid 10 grs.; water 1 oz. *British Skin.*
Carbolic Acid 24 grs.; water 1 oz. *Guy's.*
Carbolic Acid 1 oz.; water to 40 oz. *Middlesex. Royal Free. St. George's. Women.*
Carbolic Acid Crystals 1 gr.; water to 55 mins. *Fever.*
Carbolic Acid 1 gr.; water to 80 mins. *St. Thomas's.*
Carbolic Acid Crystals 1 gr.; Mucilage 4 mins.; water to 80 mins. *King's.*
Carbolic Acid 1 oz.; Glycerine 1 oz.; water 40 oz. *Charing Cross.*
Glycerine of Carbolic Acid 15 mins.; water to 1 oz. *Westminster.*
Liquefied Carbolic Acid 1 drm.; water 60 drms. *St. Mary's.*
Liquefied Carbolic Acid 50 mins.; water to 10 oz. *St. Bartholomew's*

Lotio Acidi Carbolici c. Borace.
Bicarbonate of Soda 20 grs.; Borax 20 grs.; Glycerine of Carbolic Acid 1 drm.; water to 1 oz. *St. Bartholomew's.*

Lotio Acidi Carbolici c. Borace Fortior.
Bicarbonate of Soda 120 grs.; Borax 120 grs.; Glycerine of Carbolic Acid 1½ drm.; water to 1 oz. *St. Bartholomew's.*

Lotio Acidi Carbolici c. Iodo.
Carbolic Acid Liquefied 12 mins.; Tincture of Iodine 6 mins.; water to 1 oz. *St. Bartholomew's.*

Lotio Acidi Carbolici Oleata.
Carbolic Acid 1 drm.; Olive Oil 12 drms. *Women.*

Lotio Acidi Hydrocyanici.
Diluted Hydrocyanic Acid 3 mins.; Glycerine 1 drm.; Caramel 3 mins.; water to 1 oz. *Royal Free.*
Diluted Hydrocyanic Acid 5 mins.; water 1 oz. *London Ophthalmic.*
Diluted Hydrocyanic Acid 10 mins.; water to 1 oz. *British Skin. Guy's. St. Bartholomew's.*
Diluted Hydrocyanic Acid 12 mins.; Rectified Spirit 36 mins.; water to 1 oz. *St. Thomas's.*
Diluted Hydrocyanic Acid 30 mins.; water to 1 oz. *St. George's.*

Lotio Acidi Hydrocyanici c. Sodâ.
Diluted Hydrocyanic Acid 5 mins.; Bicarbonate of Soda 5 grs.; Borax 5 grs.; water 1 oz. *London Ophthalmic.*

Lotio Acidi Nitrici.
Diluted Nitric Acid 6 mins.; water to 1 oz. *London.*
Diluted Nitric Acid 7½ mins.; water to 1 oz. *Westminster.*
Diluted Nitric Acid 10 mins.; water 1 oz. *British Skin. Guy's.*
Diluted Nitric Acid 15 mins.; Tincture of Opium 15 mins.; water to 1 oz. *University.*
Diluted Nitric Acid 20 mins.; water to 1 oz. *Royal Free.*
Strong Nitric Acid 2 mins.; water 1 oz. *Gt. Northern.*
Strong Nitric Acid 3 mins.; water to 1 oz. *St. Thomas's.*
Strong Nitric Acid 4 mins.; Tincture of Opium 6 mins.; water 1 oz. *Fever.*
Strong Nitric Acid 35 mins.; water 12 oz. *St. Mary's.*

Lotio Acidi Nitrici et Opii.

Dil. Nitric Acid 5 mins.; Tincture of Opium 5 mins.; water to 1 oz. *Westminster.*
Dil. Nitric Acid 15 mins.; Liquid Extract of Opium 45 mins; water to 2 oz. *Middlesex.*
Diluted Nitric Acid 3 mins.; Opium Lotion to 1 oz. *St. Bartholomew's.*
Nitric Acid 3 mins.; Tincture of Opium 3 mins.; water to 1 oz. *Charing Cross.*

Lotio Acidi Nitro-Hydrochlor.

Nitric Acid 3 mins.; Hydrochloric Acid 6 mins.; water 1 oz. *St. Thomas's.*

Lotio Acidi Sulphurosi.

Sulphurous Acid 1 drm.; water to 1 oz. *Middlesex.*
Sulphurous Acid 2 drms.; water to 1 oz. *British Skin. Charing Cross. London.*
Sulphurous Acid 5 oz.; Glycerine to 10 oz. *Westminster.*

Lotio Acidi Sulphurosi Fortior.

The "Sulphurous Acid." B.P. *London*

Lotio Acidi Tannici.

Glycerine of Tannic Acid 1 oz.; water to 20 oz. *Charing Cross.*
Tannic Acid 3 grs.; Water to 1 oz. *Royal Free.*
Tannic Acid 10 grs.; water 1 oz. *Guy's.*

Lotio Alkalina.

Bicarbonate of Potash 4 drms.; water 12 oz. *St. Mary's.*
Carbonate of Potash ½ oz.; Iodide of Potassium 60 grs.; tepid water 20 oz. *Charing Cross.*
Bicarbonate of Soda 12 grs.; Carbolic Acid 1½ gr.; water 1 oz. *Throat.*
Carbonate of Soda 7 grs.; water 1 oz. *Guy's.*

Lotio Aluminis.

Alum 3 grs.; water 1 oz. *Westminster Ophthalmic.* No. 1.
Alum 4 grs.; water to 1 oz. *St. Bartholomew's.*
Alum 4½ grs.; water 1 oz. *St. Thomas's.*
Alum 6 grs.; water 1 oz. *St. Mary's. Gt. Northern. London Ophthalmic. Westminster Ophthalmic.* No. 2.

Alum 10 grs. ; water 1 oz. *British Skin. Guy's. London.*
Alum 12 grs.; water 1 oz. *Charing Cross. Westminster.*
Alum 15 grs. ; water 1 oz. *City Chest.*
Alum 6 grs.; Carbolic Acid 1½ gr.; water 1 oz. *Throat.*

Lotio Aluminis Ætherea.
Alum 6 grs. ; Spirit of Nitrous Ether 1 drm.; water to 1 oz. *St. Bartholomew's.*

Lotio Aluminis et Acidi Tannici.
Alum 12 grs.; Tannic Acid 6 grs.; water 1 oz. *Westminster.*

Lotio Aluminis Co.
Alum 8 grs.; Sulphate of Zinc 8 grs.; water to 1 oz. *St. Bartholomew's.*
Alum 3 grs.; Sulphate of Zinc 1 gr.; Distilled Water 1 oz. *Westminster Ophthalmic. London Ophthalmic.*

Lotio Aluminis c. Decoct. Quercûs.
Alum 4½ grs.; Decoction of Oak Bark 1 oz. *St. Thomas's.*

Lotio Aluminis Ferri.
Iron Alum 6 grs.; water 1 oz. *St. Mary's.*

Lotio Aluminis Fortior.
Alum 16 grs. ; water to 1 oz. *St. Bartholomew's.*

Lotio Aluminis Mitis.
Alum 3 grs. ; water 1 oz. *London Ophthalmic.*
Alum 6 grs. ; water to 1 oz. *Westminster.*

Lotio Aluminis et Zinci.
Alum 4 grs.; Sulphate of Zinc 1 gr.; water 1 oz. *Westminster.*

Lotio Aluminis et Zinci Sulphatis.
Alum 10 grs.; Sulphate of Zinc 10 grs.; water 1 oz. *Guy's.*

Lotio Ammoniæ Acetatis.
Solution of Acetate of Ammonia 1 oz.; water 9 oz. *Westminster Ophthalmic.*
Solution of Acetate of Ammonia (strong) 1 part; water 10 parts. *St. George's.*

Lotio Ammonii Chloridi.
> Chloride of Ammonium 15 grs.; Diluted Acetic Acid ½ drm.; Rectified Spirit ½ drm.; water to 1 oz. *London.*
> Chloride of Ammonium 30 grs.; Distilled Vinegar 1 drm.; Spirit of Wine 1 drm.; water to 1 oz. *Fever. St. George's.*

Lotio Ammonii Chloridi Alkalina.
> Chloride of Ammonium 6 grs.; Bicarbonate of Soda 6 grs.; water 1 oz. *Throat.*

Lotio Ammonii Chloridi Astringens.
> Chloride of Ammonium 6 grs.; Alum 6 grs.; water 1 oz. *Throat.*

Lotio Anthemidis.
> Chamomile Flowers 30 grs.; Boiling Distilled Water 1 oz. Infuse 15 minutes and strain. *British Skin.*

Lotio Argenti Nitratis.
> Nitrate of Silver ½ gr.; Distilled Water 1 oz. *St. Mary's.*
> Nitrate of Silver 1 gr.; Distilled Water 1 oz. *St. Bartholomew's.*
> Nitrate of Silver 2 grs.; Distilled Water to 1 oz. *Westminster.*
> Nitrate of Silver 5 grs.; water 1 oz. *Guy's.*
> Nitrate of Silver 3 grs.; Distilled Water to 1 oz. *British Skin. London.*

Lotio Argenti Nitratis Ætherea.
> Nitrate of Silver 2 grs.; Spirit of Nitrous Ether ½ drm.; Distilled Water to 1 oz. *St. Bartholomew's.*

Lotio Argenti Nitratis Fort.
> Nitrate of Silver 1 drm.; Distilled Water 1 oz. *Fever.*
> Nitrate of Silver 2 grs.; Distilled Water 1 oz. *St. Bartholomew's.*

Lotio Arnicæ.
> Tincture of Arnica 20 mins.; water 1 oz. *London Ophthalmic.*

Tincture of Arnica 24 mins.; water to 1 oz. *St. George's.*
Tincture of Arnica 1 drm.; Lead Lotion to 1 oz. *St. Bartholomew's.*

Lotio Atropiæ et Hydrargyri Perchloridi.
Atropia Lotion 4 drms.; Perchloride of Mercury Lotion 4 drms.: mix. (Eye Lotion No. 3.) *University.*

Lotio Atropiæ Sulphatis.
Solution of Sulphate of Atropia 12 mins.; water to 1 oz. (Eye Lotion No. 2.) *University.*
Sulphate of Atropia 2 grs.; water 1 oz. *St. Bartholomew's.*

Lotio Belladonnæ.
Extract of Belladonna 2 grs.; water 1 oz. *St. Mary's.*

Lotio Belladonnæ c. Alumine.
Extract of Belladonna 5 grs.; Alum 3 grs.; water 1 oz. *Gt. Northern. London Ophthalmic.*

Lotio Bismuthi.
Subnitrate of Bismuth 10 grs.; water 1 oz. *British Skin.*

Lotio Bismuthi et Hydrargyri.
Subnitrate of Bismuth 5 grs.; Perchloride of Mercury $\frac{1}{5}$ gr.; Oxide of Zinc 20 grs.; water 1 oz. *University.*

Lotio Bismuthi et Zinci.
Oxide of Bismuth 20 grs.; Oxide of Zinc 20 grs.; Glycerine $\frac{1}{2}$ drm.; water to 1 oz. *London.*

Lotio Bituminis Comp.
Compound Solution of Bitumen (Hosp.) 20 mins.; water to 1 oz. *London.*

Lotio Boracis.
Borax 3 grs.; water 1 oz. *Samaritan.*
Borax $4\frac{1}{2}$ grs.; water to 1 oz. *St. Thomas's.*
Borax 10 grs.; water 1 oz. *Guy's. London.*
Borax $\frac{1}{4}$ oz.; water to 10 oz. *Middlesex.*

Borax 15 grs.; water 1 oz. *British Skin.*
Borax 6 grs.; Chlorate of Potash 5 grs.; water 1 oz. *St. Mary's.*
Borax 15 grs.; Glycerine 30 mins.; water 1 oz. *Westminster.*
Borax 36 grs.; Glycerine 72 mins.; water to 1 oz. *King's.*
Glycerine of Borax 1 drm.; water to 1 oz. *Gt. Northern. Charing Cross. London Ophthalmic.*

Lotio Boracis Composita.
Borax 40 grs.; Diluted Hydrocyanic Acid 6 mins.; water 10 oz. *Middlesex.*

Lotio Boracis c. Glycerino.
Borax ¾ oz.; Glycerine 1½ oz.; water to 10 oz. *Royal Free.*

Lotio Boracis c. Potassæ Chloratâ.
Borax 3 grs.; Chlorate of Potash 6 grs.; water 1 oz. *St. Bartholomew's.*

Lotio Boracis c. Zinco.
Borax 3 grs.; Sulphate of Zinc 4 grs.; Tincture of Opium 5 mins.; Glycerine 1 drm.; water to 1 oz. *St. Bartholomew's.*
Borax 4 grs.; Sulphate of Zinc 4 grs.; Glycerine 1 drm.; water to 1 oz. *Gt. Northern.*

Lotio Calaminæ.
Calamine 30 grs.; Oxide of Zinc 20 grs.; Glycerine ½ drm.; water to 1 oz. *London.*
Calamine 30 grs.; Oxide of Zinc 20 grs.; water 1 oz. *Middlesex.*
Levigated Calamine 6 drms.; Oxide of Zinc 6 drms.; Glycerine 2 oz.; Lime Water to 12 oz. *St. Mary's.*
Levigated Calamine 40 grs.; Oxide of Zinc 20 grs.; Glycerine 20 mins.; water to 1 oz. *University.*
Prepared Calamine 1½ oz.; Oxide of Zinc ½ oz.; Glycerine 1 oz.; Lime Water 8 oz.; water 4 oz. *Charing Cross.*

Lotio Calcis.
Lime Water 1 oz.; Almond Oil 1 drm. *British Skin.*

Lotio Calcis Chloratæ.
> Solution of Chlorinated Lime 3 mins.; water to 1 oz. *London.*

Lotio Calcis et Sodæ.
> Bicarbonate of Soda 6 grs.: Glycerine 10 mins.; Lime Water to 1 oz. *Westminster.*

Lotio Calcii Sulphurati.
> Slaked Lime 4 oz.; Sublimed Sulphur 4 oz.; water 35 oz.; boil and evaporate to 1 pint and filter. *University.*

Lotio Calcis et Sulphuris.
> Sulphur 12 grs.; Lime Water 1 oz. *Westminster.*

Lotio Carbonis.
> Liq. Carbonis Deterg. 4 drms.; water 20 oz. *Women.*

Lotio Carbonis Detergens.
> Liq. Carbonis. Deterg. 4 drms.; water to 12 oz. *St. Mary's.*
> Liq. Carbonis Detergens ½ to 1 fl. oz.; water to 20 oz. *Guy's.*
> Liq. Carbonis Detergens ½ oz.; Glycerine ½ oz.; water to 20 oz. *Charing Cross.*

Lotio Chloroformi.
> Chloroform 2 drms.; Almond Oil 2 oz. *Women.*

Lotio Conii.
> Extract of Conium 120 grs.; Diluted Solution of Subacetate of Lead 12 oz. *St. George's.*

Lotio Cupri Sulphatis.
> Sulphate of Copper 2 grs.; water to 1 oz. *St. Bartholomew's.*
> Sulphate of Copper 5 grs.; water to 1 oz. *Royal Free.*
> Sulphate of Copper 3 grs.; water 1 oz. *British Skin.*

Lotio Evaporans.
> Chloride of Ammonium ½ oz.; Spirit of Wine (methylated) 2 oz.; water to 20 oz. *Charing Cross.*
> Methylated Spirit 1½ oz.; water to 20 oz. *Royal Free.*

Solution of Acetate of Ammonia 108 mins.; Spirit
of Wine 108 mins.; water to 1 oz. *St. Thomas's.*
Rectified Spirit 2 drms.; water to 1 oz. *University.*
Solution of Acetate of Ammonia 1 drm.; Rectified
Spirit 2 drms.; water to 1 oz. *Westminster.*
Hydrochlorate of Ammonia 12 grs.; Spirit of Wine
36 mins.; water to 1 oz. *King's.*

Lotio Ferri Sulphatis.
Sulphate of Iron 3 grs.; water 1 oz. *St. Bartholomew's.*
Sulphate of Iron 5 grs.; water to 1 oz. *Middlesex. London Ophthalmic.*

Lotio Ferri Tartarati.
Tartarated Iron 30 grs.; water 1 oz. *British Skin.*

Lotio Flava.
Corrosive Sublimate 1 gr.; Lime Water 1 oz. *St. George's.*

Lotio Frigida.
Solution of Acetate of Ammonia 3½ drms.; water to 10 drms. *St. Thomas's.*

Lotio Glycerini.
Glycerine 1 drm.; water to 1 oz. *British Skin.*
Glycerine ½ oz.; Saccharated Solution of Lime ½ oz. *Samaritan.*

Lotio Glycerini Alkalina.
Glycerine 1 fl. drm.; Carbonate of Soda 5 grs.; water to 1 oz. *Guy's.*

Lotio Glycerini Carbolici.
Glycerine of Carbolic Acid 3 drms.; water to 12 oz. *St. Mary's.*

Lotio Glycerini et Rosæ.
Glycerine and Rose Water equal parts. *City Chest.*

Lotio Hydr. Cinerea. *See* Lotio Nigra.

Lotio Hydr. Flava. *See* Lotio Flava.

Lotio Hydr. Nigra. *See* Lotio Nigra.

Lotio Hydr. Oxyd. *See* Lotio Nigra.

Lotio Hydrarg. Perchlor.
Corrosive Sublimate ⅛ gr.; water 1 oz. *London. Ophthalmic.*
Corrosive Sublimate ¼ gr.; water 1 oz. *St. Bartholomew's.*
Corrosive Sublimate 1 gr.; water to 1 oz. *London. British Skin.*
Corrosive Sublimate 1/10 gr.; Tincture of Cochineal 3 mins.; water to 1 oz. (Eye Lotion No. 1.) *University.*
Corrosive Sublimate 2 grs.; water 1 oz. *Guy's.*

Lotio Hydrarg. Perchlor. Hydrocyanica.
Corrosive Sublimate ½ gr.; Glycerine 2 drms.; Diluted Hydrocyanic Acid 8 mins.; water to 1 oz. *St. Bartholomew's.*

Lotio Hydr. Subchlor. *See* Lotio Nigra.

Lotio Inulæ.
Elecampane Root sliced 15 grs.; water 2 oz. Boil down to 1 oz. and strain. *British Skin.*

Lotio Iodi.
Tincture of Iodine 3 mins.; water to 1 oz. *St. Thomas's.*
Tincture of Iodine 12 mins.; water 1 oz. *Fever. Women.*
Tincture of Iodine 20 mins.; water to 1 oz. *Guy's. London.*
Methylated Tincture of Iodine 1 oz.; water to 20 oz. *Royal Free.*

Lotio Iodi Co.
Iodine 60 grs.; Iodide of Potassium 80 grs.; water 1 oz. *St. George's.*

Lotio Myrrhæ.
Tincture of Myrrh 72 mins.; water to 1 oz. *King's.*

Lotio Nigra.
Calomel 30 grs.; Mucilage of Acacia 2 drms.; Lime Water to 10 oz. *Middlesex.*
Calomel 15 grs.; Mucilage 7½ mins.; Lime Water 1 oz. *St. George's.*
Calomel 3 grs.; Solution of Potash 3 mins.; water 1 oz. *Samaritan.*

Calomel 10 grs.; Mucilage of Tragacanth 1 drm.; Lime Water 1 oz. *British Skin.*

Lotio Opii.
Extract of Opium 4 grs.; water 1 oz. *London.*
Extract of Opium 5 grs.; water 1 oz *Guy's.*
Extract of Opium 6 grs.; warm water to 1 oz. *St. Thomas's.*
Opium 5 grs.; boiling water to 1 oz. *St. Bartholomew's.*

Lotio Picis Ligni Fossilis.
(Syn. Coal Tar Emulsion.)
Coal Tar Solution * 2 drms.; water 1 oz. Mix to form an emulsion. Diluted most commonly with six or twelve times its quantity of water as a mild stimulant in cases of Chronic Eczema and of Seborrhœa. *British Skin.*

Lotio Plumbi.
The same as Liquor Plumbi Subacetatis Dilutus, Pharm. Brit. *Fever. London. Westminster.*
Solution of Subacetate of Lead 1 oz.; water to 40 oz. *Royal Free.*
Solution of Subacetate of Lead 6 mins.; water to 1 oz. *St. Bartholomew's.*
Solution of Subacetate of Lead 5 mins.; Rectified Spirit 5 mins; water 1 oz. *Gt. Northern.*
Acetate of Lead 16 grs.; Dil. Acetic Acid 16 mins.; Distilled Water 10 oz. *Westminster Ophthalmic.*

Lotio Plumbi Acetatis.
Acetate of Lead 2 grs.; Diluted Acetic Acid 2 mins.; water to 1 oz. *London Ophthalmic.*
Acetate of Lead 3 grs.; Glycerine 36 mins.; water to 1 oz. *King's.*
Acetate of Lead 2 grs.; Acetic Acid 2 mins.; water 1 oz. *Westminster.*

Lotio Plumbi Acetatis c. Morphiâ.
Acetate of Lead 1 gr.; Acetate of Morphia 1 gr.; water 1 oz. *Guy's.*

Lotio Plumbi Co.
Solution of Subacetate of Lead 1 oz.; Diluted Acetic Acid 1 oz.; Tincture of Opium 2 oz. A Dessert-spoonful to 20 oz. tepid water. *Samaritan.*

---

* *See* Pigmentum Picis Ligni Fossilis.

Lotio Plumbi Evaporans.
Solution of Subacetate of Lead 6 mins.; Rectified Spirit 2 drms.; Distilled Water to 1 oz. *London.*

Lotio Plumbi Evaporans Fortior.
Solution of Subacetate of Lead 1 drm.; Rectified Spirit 4 drms.; Distilled Water to 1 oz. *London.*

Lotio Plumbi Fortior.
Solution of Subacetate of Lead ½ oz.; Rectified Spirit (methylated) ½ oz.; water to 20 oz. *Charing Cross.*
Solution of Subacetate of Lead 16 mins.; Rectified Spirit 16 mins.; water to 1 oz. *London Ophthalmic.*

Lotio Plumbi c. Opio.
Acetate of Lead 4 grs.; Opium in powder 4 grs.; warm water 1 oz. *Fever.*
Acetate of Lead 4 grs.; Tincture of Opium 4 mins.; water 1 oz. *St. Mary's.*
Acetate of Lead 4 grs.; Extract of Opium 1 gr.; hot water 1 oz. *Women.*
Extract of Opium 1 gr.; Diluted Solution of Subacetate of Lead to 1 oz. *Westminster.*
Extract of Opium 4 grs.; Diluted Solution of Subacetate of Lead 1 oz. *London.*
Methylated Tincture of Opium 1 oz.; Lead Lotion to 20 oz. *Royal Free.*
Opium Lotion and Lead Lotion equal parts. *St. Bartholomew's. Guy's.*
Opium 15 grs.; Diluted Solution of Subacetate of Lead 10 oz. *Middlesex.*

Lotio Plumbi Subacetatis.
Solution of Subacetate of Lead 5 mins.; water 1 oz. *British Skin.*

Lotio Potassæ Carbonatis.
Carbonate of Potash 24 grs.; water 1 oz. *King's.*

Lotio Potassæ Chloratis.
Chlorate of Potash 10 grs.; water 1 oz. *Guy's. London.*
Chlorate of Potash 27 grs.; water 1 oz. *British Skin.*

Chlorate of Potash 5 grs.; Compound Spirit of
Lavender 2½ mins.; water 1 oz. *Samaritan.*
Chlorate of Potash 5 grs.; Diluted Hydrochloric
Acid 2 mins.; water 1 oz. *London Ophthalmic.*

Lotio Potassæ Chloratis Alkalina.
Chlorate of Potash 6 grs.; Bicarbonate of Soda
6 grs.; water 1 oz. *Throat.*

Lotio Potassæ Permanganatis.
Permanganate of Potash 6 grs.; water 20 oz.
Charing Cross.
Solution of Permanganate of Potash 40 mins.;
water 1 oz. *London.*
Solution of Permanganate of Potash 12 mins.;
water 1 oz. *Fever. Middlesex. University. Westminster.* (*Women* same as Injection.)
Solution of Permanganate of Potash 10 mins.;
water 1 oz. *Guy's.*

Lotio Potassæ Sulphuratæ.
Sulphurated Potash 4 grs.; water 1 oz. *St. Bartholomew's.*
Sulphurated Potash 3¾ grs.; Lime Water 1 oz.
St. George's.
Sulphurated Potash 9 grs.; water to 1 oz. *Middlesex.*
Sulphurated Potash 10 grs.; water 1 oz. *British Skin.*

Lotio Quercûs Co.
Alum 4 grs.; Sulphate of Zinc 4 grs.; Decoction of
Oak Bark to 1 oz. *St. Bartholomew's.*

Lotio Resorcini. (Syn. Andeers Lotion.)
Resorcin 40 grs.; water 1 oz. Dissolve. An antiseptic and stimulant in foul or syphilitic ulcerations and to allay irritation in chronic Eczema
and Psoriasis. *British Skin.*

Lotio Rubra.
Sulphate of Zinc 1 gr.; Spirit of Rosemary 15
mins.; Compound Tincture of Lavender 15
mins.; water to 1 oz. *London Ophthalmic.
Middlesex.*
Sulphate of Zinc 1 drm.; Colouring 10 drops;
water to 20 oz. *Royal Free.*
Sulphate of Zinc 2 grs.; Comp. Tincture of Lavender 12 mins.; water to 1 oz. *University.*

Sulphate of Zinc 2 grs.; Compound Tincture of
Lavender 10 mins.; water 1 oz. *London. St.
Mary's.*
Sulphate of Zinc 1 gr.; Spirit of Rosemary 9 mins.;
Comp. Spirit of Lavender 9 mins.; water to 1
oz. *King's.*
Perchloride of Mercury 1 gr.; Bisulphuret of Mercury $\frac{1}{2}$ gr.; Creasote $\frac{1}{2}$ min.; water 1 oz.
Samaritan.
Sulphate of Copper 60 grs.; Camphor 15 grs.; Armen.
Bole 60 grs.; boiling water 16 oz. *St. George's.*

Lotio Sodæ Bicarbonatis.
Bicarbonate of Soda 4 grs.; water 1 oz. *Westminster
Ophthalmic.*

Lotio Sodæ Bicarbonatis et Opii.
Bicarbonate of Soda 8 grs.; Tincture of Opium 8
mins.; water 1 oz. *Westminster.*

Lotio Sodæ Carbonatis.
Carbonate of Soda 10 grs.; water 1 oz. *London.*

Lotio Sodæ Chloratæ.
Chlorinated Solution of Soda 5 mins.; water to 1 oz.
Westminster.
Chlorinated Solution of Soda 30 mins.; water to 1
oz. *British Skin. St. Bartholomew's. St.
Mary's.*
Chlorinated Solution of Soda 40 mins.; water to 1
oz. *London.*
Chlorinated Solution of Soda 48 mins.; water to 1
oz. *Fever.*
Chlorinated Solution of Soda 60 mins.; water 1
oz. *Guy's.*
Chlorinated Solution of Soda 84 mins.; water to 1
oz. *St. Thomas's.*

Lotio Sodæ Hyposulphitis.
Hyposulphite of Soda 1 drm.; water 1 oz. *City
Chest. Guy's.*
A saturated solution in water. *King's.*
Hyposulphite of Soda 9 grs.; Sulphurous Acid 36
mins.; water to 1 oz. *Samaritan.*
Hyposulphite of Soda 3 oz.; Glycerine 1 oz.; water
to 20 oz. *Charing Cross.*
Hyposulphite of Soda 15 grs.; Sulphurous Acid 36
mins.; water to 1 oz. *Middlesex.*

Lotio Spirituosa.
Rectified Spirit 1 oz.; water to 10 oz. *Middlesex.*

Lotio Spiritus.
Rectified Spirit 2 drms.; water to 1 oz. *Gt. Northern. Guy's. London.*

Lotio Stimulans.
Tincture of Cantharides 5 mins.; Stronger Solution of Ammonia 10 mins.; Rectified Spirit 10 mins.; Glycerine 10 mins.; Oil of Rosemary 3 mins.; water to 1 oz. *Westminster.*

Lotio Sulphuris.
Precipitated Sulphur 10 drms.; Rectified Spirit 5 oz.; water to 10 oz. *Middlesex.*

Lotio Sulphuris Co.
Precipitated Sulphur 10 drms.; Spirit of Camphor 5 drms.; Lime Water to 10 oz. *St. Bartholomew's.*

Lotio Tabaci c. Plumbo.
Solution of Subacetate of Lead 3 mins.; Infusion of Tobacco 1 oz. *Women.*

Lotio Thymol (1 in 1000).
Thymol 10 grs.; Hot Water 20 oz.; Dissolve. *Guy's. London.*

Lotio Vaselini.
Vaseline 1 oz.; Solution of Subacetate of Lead ½ oz. *Samaritan.*

Lotio Zinci.
Sulphate of Zinc 2 grs.; water to 1 oz. *Royal Chest.*

Lotio Zinci et Acidi Carbolici.
Sulphate of Zinc 5 grs.; Carbolic Acid 5 mins.; Glycerine 30 mins.; water to 1 oz. *Westminster.*

Lotio Zinci Chlor.
Chloride of Zinc 2 grs.; water 1 oz. *Women.*
Chloride of Zinc 10 grs.; water 10 oz. *Charing Cross. Middlesex. St. Mary's.*

Chloride of Zinc 4 grs.; water 1 oz. *Westminster.*
Solution of Chloride of Zinc 1½ min.; water to 1 oz. *St. Thomas's.*

Lotio Zinci c. Glycerinâ.

Oxide of Zinc ½ oz.; Glycerine ½ oz.; water to 6 oz. *Samaritan.*

Lotio Zinci Oxydi.

Oxide of Zinc 20 grs.; Glycerine 1 oz.; water to 10 oz. *Royal Free.*
Oxide of Zinc 24 grs.; Glycerine ½ drm.; water to 1 oz. *London.*
Oxide of Zinc 30 grs.; Glycerine 60 mins.; water to 1 oz. *Guy's.*
Oxide of Zinc 1 oz.; Lime Water 20 oz. *Charing Cross.*
Oxide of Zinc 30 grs.; Glycerine ½ drm.; Lime Water to 1 oz. *St. Bartholomew's.*
Oxide of Zinc 60 grs.; Prepared Calamine 60 grs.; Glycerine 1 drm.; water 1 oz. *British Skin.*

Lotio Zinci Sulphatis.

Sulphate of Zinc 1 gr.; water to 1 oz. *London Ophthalmic. St. Bartholomew's. Westminster Ophthalmic.*
Sulphate of Zinc 2 grs.; Burnt Sugar 3 grs.; water 1 oz. *Charing Cross.*
Sulphate of Zinc 6 grs.; Carbolic Acid 1½ gr.; water 1 oz. *Throat.*
Sulphate of Zinc 3 grs.; water 1 oz. *Royal Free.*
Sulphate of Zinc 2 grs.; water 1 oz. *Guy's. St. Mary's. St. Thomas's. Westminster.*
Sulphate of Zinc 1 gr.; Comp. Tincture of Lavender 18 mins.; water to 1 oz. *Fever.*
Sulphate of Zinc 5 grs.; water 1 oz. *London. Middlesex.*
Sulphate of Zinc 2 grs.; Compound Tincture of Lavender 25 mins.; water to 1 oz. *British Skin.*

Lotio Zinci Sulphatis et Acidi Carbolici.

Sulphate of Zinc 2 grs.; Carbolic Acid 10 grs.; water to 1 oz. *Guy's.*

Lotio Zinci Sulphatis Fortior.

Sulphate of Zinc 40 grs.; water 10 oz. *St. Bartholomew's.*

Lotio Zinci Sulphocarbolatis.
Sulphocarbolate of Zinc 40 grs.; water to 10 oz. *St. Bartholomew's.*
Sulphocarbolate of Zinc 50 grs.; water 10 oz. *London.*

MASSA PRO PESSIS. See Pessus.

MISTURÆ.

Mistura Acaciæ.
Acacia Gum 150 grs.; water to 1 oz. *Throat.*
Gum Acacia 40 grs.; water to 1 oz. *London.*

Mistura Acaciæ Opiata.
Comp. Tincture of Camphor ½ drm.; Chloric Ether 5 mins.; Mucilage 2 drms.; water to 1 oz. *Consumption.*

Mistura Acida.
Dil. Nitric Acid 10 mins.; Dil. Hydrochloric Acid 8 mins.; Comp. Infusion of Gentian to 1 oz. *Royal Free.*
Dil. Nitro-Hydrochloric Acid 10 mins.; Infusion of Quassia 1 oz. *Gt. Northern.*

Mistura Acida Amara.
Dil. Nitro-Hydrochloric Acid 10 mins.; Comp. Infusion of Gentian 1 oz. *St. Mary's.*

Mistura Acida Aromatica.
Dil. Hydrochloric Acid 15 mins.; Comp. Spirit of Horseradish 10 mins.; Tincture of Ginger 5 mins.; Infusion of Calumba to 1 oz. *King's.*
Alcoholized Sulphuric Acid 12 mins.; water to 1 oz. *Throat.*

Mistura Acida Digitalis.
Dil. Sulphuric Acid 15 mins.; Tincture of Digitalis 10 mins.; Infusion of Quassia to 1 oz. *Royal Chest.*

Mistura Acida c. Opio.
Dil. Sulphuric Acid 20 mins.; Tincture of Opium 5 mins.; Spirit of Chloroform 20 mins.; Cinnamon Water to 1 oz. *Charing Cross.*

Mistura Acida c. Quassiâ.
Dil. Nitro-Hydrochloric Acid 10 mins.; Infusion of Quassia 1 oz. *Consumption.*

Mistura Acidi Acetici et Morphiæ.
Solution of Acetate of Morphia 20 mins.; Dil. Acetic Acid 30 mins.; Camphor Water to 1 oz. *Royal Chest.*

Mistura Acidi Benzoici Ammoniata.
Benzoic Acid 5 grs.; Solution of Ammonia 5 mins.; water to 1 oz. *London.*

Mistura Acidi Composita.
Tincture of Nux Vomica 10 mins.; Comp. Tincture of Gentian 60 mins.; Mixture of Nitro-Hydrochloric Acid to 1 oz. *Guy's.*

Mistura Acidi Fluorici.
Diluted Fluoric Acid ($\frac{1}{2}$ per cent. solution of the redistilled acid) 10 mins.; water 1 oz. *Throat.*

Mistura Acidi Gallici.
Gallic Acid 10 grs.; Tincture of Opium 3 mins.; Dil. Sulphuric Acid 15 mins.; water to 1 oz. *University.*
Gallic Acid 10 grs.; Comp. Tincture of Camphor 20 mins.; Dil. Sulphuric Acid 15 mins.; water to 1 oz. *Consumption.*
Gallic Acid 10 grs.; Glycerine 40 mins.; water to 1 oz. For a dose. *Throat.*
Glycerine of Gallic Acid $\frac{1}{2}$ drm.; Dil. Sulphuric Acid 15 mins.; water 1 oz. *Royal Chest.*
Glycerine of Gallic Acid $\frac{1}{2}$ drm.; Glycerine $\frac{1}{2}$ drm.; water to 1 oz. *Royal Free.*

Mistura Acidi Gallici c. Opio.
Gallic Acid 10 grs.; Dil. Sulphuric Acid 10 mins.; Tincture of Opium 5 mins.; water to 1 oz. *London.*
Gallic Acid 10 grs.; Dil. Sulphuric Acid 20 mins.; Comp. Tincture of Camphor 20 mins.; water to 1 oz. *Charing Cross.*

Mistura Acidi Hydrobromici.
Diluted Hydrobromic Acid 15 mins.; water to 1 oz. *Throat.*

Mistura Acidi Hydrochlorici.

Dil. Hydrochloric Acid 10 mins.; Simple Syrup 60 mins.; water to 1 oz. *Guy's.*

Dil. Hydrochloric Acid 10 mins.; Tincture of Chiretta 15 mins.; water to 1 oz. *London Ophthalmic.*

Mistura Acidi Hydrocyanici.

Dil. Hydrocyanic Acid 5 mins.; Bicarbonate of Soda 10 grs.; water 1 oz. *Gt. Northern.*

Dil. Hydrocyanic Acid 3 mins.; Caramel 5 mins.; water to 1 oz. *Royal Free.*

Dil. Hydrocyanic Acid 3 mins.; Red Mixture to 1 oz. *London.*

Dil. Hydrocyanic Acid 5 mins.; Bicarbonate of Soda 20 grs.; water 1 oz. *St. Mary's.*

Dil. Hydrocyanic Acid 4 mins.; Bicarbonate of Soda 15 grs.; water to 1 oz. *Royal Chest.*

Dil. Hydrocyanic Acid 3 mins.; Camphor Water to 1 oz. *Throat.*

Dil. Hydrocyanic Acid 3 mins.; Bicarbonate of Soda 10 grs.; Tincture of Belladonna 10 mins.; Compound Infusion of Gentian 1 oz. *Samaritan.*

Mistura Acidi Hydrocyanici c. Sodâ.

Dil. Hydrocyanic Acid 4 mins.; Bicarbonate of Soda 15 grs.; water to 1 oz. *Consumption.*

Mistura Acidi Nitrici.

Dil. Nitric Acid 10 mins.; Simple Syrup 60 mins.; water to 1 oz. *Guy's.*

Dil. Nitric Acid 10 mins.; Acacia Mixture to 1 oz. *London.*

Dil. Nitric Acid 10 mins.; Infusion of Quassia to 1 oz. *Royal Free.*

Dil. Nitric Acid 20 mins.; Comp. Tincture of Cardamoms 1 drm.; Comp. Infusion of Gentian to 1½ oz. *St. Thomas's.*

Mistura Acidi Nitrici c. Calumbâ.

Dil. Nitric Acid 10 mins.; Tincture of Calumba ½ drm.; water to 1 oz. *Consumption.*

Mistura Acidi Nitrici c. Pareirâ.

Dil. Nitric Acid 10 mins.; Decoction of Pareira to 1 oz. *Royal Free.*

Mistura Acidi Nitrici c. Strychniâ.

Solution of Strychnia 5 mins.; Nitric Acid Mixture to 1 oz. *London.*

Mistura Acidi Nitro-Hydrochlorici.

Dil. Nitro-Muriatic Acid 10 mins.; Simple Syrup 60 mins.; water to 1 oz. *Guy's.*
Dil. Nitro-Muriatic Acid 10 mins.; Red Mixture to 1 oz. *London.*
Dil. Nitro-Hydrochloric Acid 15 mins.; Syrup of Orange ½ drm.; water to 1 oz. *Consumption.*
Dil. Nitro-Hydrochloric Acid 10 mins.; Infusion of Quassia to 1 oz. For a dose. *Throat.*
Dil. Nitric Acid 7 mins.; Dil. Hydrochloric Acid 8 mins.; Chloroform Water 2 drms.; Infusion of Gentian to 1 oz. *Charing Cross.*
Dil. Nitro-Hydrochloric Acid 15 mins.; Caramel 5 grs.; water 1 oz. *British Skin.*
Dil. Nitric Acid 10 mins.; Dil. Hydrochloric Acid 10 mins.; Infusion of Quassia 1 oz. *Westminster Ophthalmic.*
Dil. Nitro-Hydrochloric Acid 15 mins.; Infusion of Chiretta to 1 oz. *King's.*
Dil. Nitro-Hydrochloric Acid 15 mins.; Infusion of Quassia to 1 oz. *Royal Chest. Royal Free.*

Mistura Acidi Phosphorici.

Dil. Phosphoric Acid 15 mins.; Spirit of Chloroform 10 mins.; Comp. Infusion of Gentian to 1 oz. *University.*

Mistura Acidi Phosphorici Co.

Dil. Phosphoric Acid 15 mins.; Solution of Strychnia (B.P.) 5 mins.; Chloroform Water 2 drms.; Infusion of Quassia to 1 oz. For a dose. *Throat.*

Mistura Acidi Phosph. c. Quassiâ.

Dil. Phosphoric Acid 15 mins.; Tincture of Quassia 30 mins.; Syrup of Orange 20 mins.; water to 1 oz. *Consumption.*

Mistura Acidi Salicylici.

Salicylic Acid 20 grs.; Citrate of Potash 20 grs.; water to 1 oz. *London.*
Salicylic Acid 20 grs.; Bicarbonate of Soda 20 grs.; water 1 oz. *St. Mary's.*

Mistura Acidi Sulphurici.

Dil. Sulphuric Acid 12 mins.; Tincture of Catechu 15 mins.; water to 1 oz. *University.*
Dil. Sulphuric Acid 20 mins.; Chloroform Water ½ oz.; water to 1 oz. *Charing Cross.*

Mistura Acidi Sulphurici c. Æthere.
Dil. Sulphuric Acid 12 mins.; Spirit of Ether 10 mins.; Comp. Tincture of Camphor 12 mins.; Tinct. of Rhatany 1 drm.; Spirit of Peppermint ½ drm.; water to 1 oz. *Samaritan.*

Mistura Acidi Sulphurici Aromatici.
Aromatic Sulphuric Acid 10 mins.; Red Mixture to 1 oz. *London.*

Mistura Acidi Sulphurici Co.
Dil. Sulphuric Acid 15 mins.; Sulphate of Magnesia 20 grs.; Alum 10 grs.; water to 1 oz. *Consumption.*

Dil. Sulphuric Acid 10 mins.; Sulphate of Magnesia 40 grs.; Treacle 40 mins.; Peppermint Water 7 drms. *Westminster Ophthalmic.*

Mistura Acidi Sulphurici c. Magnesiâ.
Dil. Sulphuric Acid 10 mins.; Sulphate of Magnesia 60 grs.; Glycerine 30 mins.; Peppermint Water to 1 oz. For a dose. *London Ophthalmic.*

Mistura Acidi Sulphurici c. Opio.
Aromatic Sulphuric Acid 20 mins.; Tincture of Opium 5 mins.; water to 1 oz. *Royal Free.*

Dil. Sulphuric Acid 20 mins.; Tincture of Opium 10 mins.; Spirit of Chloroform 20 mins.; Camphor water to 1 oz. *St. Mary's.*

Dil. Sulphuric Acid 15 mins.; Tincture of Opium 6 mins.; Tinct. of Capsicum 3 mins.; water to 1 oz. *University.*

Mistura Aconiti.
Tincture of Aconite 5 mins.; Red Mixture to 1 oz. Dose 1 oz., repeated with caution. *London.*

Tincture of Aconite 1 min.; Caramel ½ min.; water to 1 oz. For a dose. *Throat.*

Mistura Aconiti et Colchici.
Tincture of Aconite 3 mins.; Colchicum Wine 10 mins.; Bicarbonate of Potash 10 grs.; water 1 oz. *Westminster Ophthalmic.*

Mistura Ætheris c. Ammoniâ.
Aromatic Spirit of Ammonia 30 mins.; Spirit of Ether 30 mins.; Tincture of Orange 10 mins.;

Camphor Water to 1 oz. *Consumption. Gt. Northern.*

Aromatic Spirit of Ammonia 20 mins.; Spirit of Ether 30 mins.; Tincture of Orange 10 mins.; Camphor Water to 1 oz. *Charing Cross.*

Mistura Ætheris et Valerianæ Ammon.
Ether ½ drm.; Ammoniated Tincture of Valerian ½ drm.; Infusion of Quassia to 1 oz. *Royal Chest.*

Mistura Æther. Chlor. c. Opio.
Chloric Ether 10 mins.; Tincture of Squill 20 mins.; Tincture of Opium 3 mins.; Honey 10 grs.; water to 1 oz. Dose ½ to 1 oz. *Consumption.*

Mistura Ætheris Co.
Spirit of Ether 1 drm.; Sal Volatile ½ drm.; Syrup ½ drm.; Pimento Water 3 drms.; water to 1 oz. 1 dose. *St. George's.*
Spirit of Ether 30 mins.; Aromatic Spirit of Ammonia 30 mins.; Syrup of Tolu 1 drm.; water 1 oz. *London Ophthalmic.*

Mistura Ætheris c. Lobeliâ.
Spirit of Ether 30 mins.; Ethereal Tincture of Lobelia 20 mins.; Camphor Water to 1 oz. Dose ½ to 1 oz. *Consumption.*

Mistura Ætheris Nitrosi. c. Ammoniâ.
Spirit of Nitrous Ether 20 mins.; Carbonate of Ammonia 3 grs.; Tincture of Tolu 10 mins.; Comp. Infusion of Gentian to 1 oz. *King's.*

Mistura Ætheris c. Sodâ.
Spirit of Ether 1 drm.; Bicarbonate of Soda 10 grs.; Peppermint Water to 1 oz. *St. Thomas's.*

Mistura Ætheris pro Tuss.
Solution of Hydrochlorate of Morphia 10 mins.; Spirit of Ether 5 mins.; water to ½ oz. 15 mins. of Comp. Tinct. of Camphor may be substituted for the Solution of Hydrochlorate of Morphia. *Samaritan.*

Mistura Alba.
Carbonate of Magnesia 6 grs.; Sulphate of Magnesia 30 grs.; Peppermint Water to 1 oz. **Throat.**

Carbonate of Magnesia 10 grs.; Sulphate of Magnesia 60 grs.; Spearmint Water 1 oz. *Gt. Northern.*

Carbonate of Magnesia 10 grs.; Sulphate of Magnesia 60 grs.; Peppermint Water 1 oz. *British Skin. King's.*

Carbonate of Magnesia 10 grs.; Sulphate of Magnesia 1½ drm.; Peppermint Water to 1 oz. *Royal Free.*

Carbonate of Magnesia 10 grs.; Sulphate of Magnesia 1 drm.; Peppermint Water ½ oz. *Samaritan.* For *St. Mary's* see Mist. Magn. Sulphatis.

Mistura Alkalina.

Carbonate of Ammonia 6 grs.; Bicarbonate of Soda 12 grs.; Infusion of Quassia to 1 oz. *Royal Free.*

Mistura Alkalina Amara.

Bicarbonate of Soda 10 grs.; Solution of Ammonia 5 mins.; Compound Infusion of Gentian 1 oz. *St. Mary's.*

Mistura Alkalina Effervescens.

Bicarbonate of Soda 15 grs.; Bicarbonate of Potash 15 grs.; water 1 oz. To be taken effervescing with 15 grs. of Tartaric Acid. *Royal Chest.*

Mistura Aloes cum Ferro.

Compound Decoction of Aloes ½ oz.; Compound Mixture of Iron ½ oz. *Royal Free.*

Mistura Alterativa.

Iodide of Potassium 5 grs.; Solution of Potash 10 mins.; Infusion of Quassia to 1 oz. *Royal Free.*

Mistura Alterativa c. Cinchonâ.

Iodide of Potassium 2½ grs.; Solution of Potash 5 mins.; Infusion of Quassia ½ oz.; Decoction of Cinchona to 1 oz. *Royal Free.*

Mistura Aluminis Co.

Alum 8 grs.; Compound Tincture of Camphor 20 mins.; Camphor Water 4 drms.; Ipecacuanha Wine 15 mins.; Dill Water to 1 oz. *Guy's.*

Mistura Amara.
Carbonate of Ammonia 5 grs.; Caramel 1 min.; Chloroform Water 2 drms.; Infusion of Quassia to 1 oz. *Throat.*

Mistura Ammoniæ.
Carbonate of Ammonia 4 grs.; Treacle 20 mins.; Comp. Tincture of Lavender 20 mins.; Peppermint Water to 1 oz. *Guy's.*

Mistura Ammoniæ Acetatis.
Solution of Acetate of Ammonia 2 drms.; water to 1 oz. *Gt. Northern. London Ophthalmic.*
Solution of Acetate of Ammonia 4 drms.; Camphor Water to 1 oz. *St. Mary's.*
Solution of Acetate of Ammonia 4 drms.; water to 1 oz. *Guy's. King's.*
Solution of Acetate of Ammonia 2 drms.; Nitrate of Potash 10 grs.; Camphor Water to 1 oz. *University.*

Mistura Ammoniæ Acetatis Co.
Solution of Acetate of Ammonia 4 drms.; Spirit of Nitrous Ether ½ drm.; Camphor Water to 1 oz. *Royal Chest.*

Mistura Ammon. Acet. c. Magnesiâ.
Sulphate of Magnesia 60 grs.; Mixture of Acetate of Ammonia 1 oz. *London Ophthalmic.*

Mistura Ammoniæ c. Æthere.
Spirit of Ether 20 mins.; Aromatic Spirit of Ammonia 20 mins.; Chloroform Water to 1 oz. *London.*
Spirit of Ether 15 mins.; Aromatic Spirit of Ammonia 15 mins.; Camphor Water to 1 oz. *University.*
Ether 20 mins.; Spirit of Chloroform 20 mins.; Aromatic Spirit of Ammonia 20 mins.; Camphor Water to 1 oz. For a dose. *Throat.*

Mistura Ammoniæ Aromatica.
Aromatic Spirit of Ammonia 15 mins.; Tincture of Cinnamon 20 mins.; Infusion of Calumba to 1 oz. *Royal Free.*

Mistura Ammoniæ Carbonatis.
Carbonate of Ammonia 5 grs.; Red Mixture 1 oz. *London.*

Mistura Ammoniæ et Chloroformi.
Aromatic Spirit of Ammonia 20 mins.; Spirit of Chloroform 20 mins.; water 1 oz. *Samaritan.*
Carbonate of Ammonia 5 grs.; Spirit of Chloroform 20 mins.; Mucilage of Acacia 2 drms.; water to 1 oz. *Royal Chest.*

Mistura Ammoniæ c. Cinchonâ.
Carbonate of Ammonia 3 grs.; Infusion of Cinchona 1 oz. *King's.*
Carbonate of Ammonia 3 grs.; Spirit of Chloroform 10 mins.; Decoction of Cinchona Bark 1 oz. *Westminster Ophthalmic.*
Carbonate of Ammonia 5 grs.; Comp. Tincture of Gentian 20 mins.; Decoction of Cinchona to 1 oz. *Royal Free.*

Mistura Ammoniæ Effervescens.
Carbonate of Ammonia 15 grs.; water to 1 oz.; with ½ oz. Tartaric Acid Solution effervescing. *London.*
Carbonate of Ammonia 15 grs.; Citric Acid 15 grs.; water to 1 oz. *Consumption.*
Carbonate of Ammonia 20 grs.; water 1 oz.; this with ½ oz. Lemon Juice. *Guy's.*
Carbonate of Ammonia 20 grs. (Tartaric Acid 25 grs., in ½ oz. of water); water to 1½ oz. *St. Mary's.*
Carbonate of Ammonia 25 grs.; water ½ oz. during effervescence with Citric Acid 20 grs.; water ½ oz. *Charing Cross.*

Mistura Ammoniæ c. Ipecac.
Carbonate of Ammonia 3 grs.; Ipecac. Wine 10 mins.; Glycerine 30 mins.; water to 1 oz. *London Ophthalmic.*
Carbonate of Ammonia 4 grs.; Ipecac. Wine 9 mins.; Camphor Water to 1 oz. *University.*
Carbonate of Ammonia 4 grs.; Ipecac. Wine 16 mins.; Treacle ¼ oz.; water to 1 oz. *Throat.*

Mistura Ammoniæ et Morphiæ.
Carbonate of Ammonia 5 grs.; Acetate of Morphia ⅛ gr.; Solution of Acetate of Ammonia 2 drms.; Camphor Water to 1 oz. *Royal Chest.*

Mistura Ammoniæ c. Quassiâ.
Carbonate of Ammonia 5 grs.; Infusion of Quassia to 1 oz. *Royal Free.*

Mistura Ammoniæ c. Senegâ.
Carbonate of Ammonia 4 grs.; Spirit of Chloroform 9 mins.; Infusion of Senega to 1 oz. *University.*
Carbonate of Ammonia 5 grs.; Infusion of Senega 1 oz. *Gt. Northern.*

Mistura Ammoniaci Co.
Comp. Tincture of Camphor 30 mins.; Oxymel of Squill 30 mins.; Mixture of Ammoniacum to 1 oz. *Consumption.*

Mistura Ammoniaci Ipecac. et Lobeliæ.
Ammoniacum Mixture ½ oz.; Ipecacuanha Wine 10 mins.; Ethereal Tincture of Lobelia 10 mins.; water to 1 oz. *Royal Chest.*

Mistura Ammonii Chloridi.
Chloride of Ammonium 10 grs.; Carbonate of Ammonia 5 grs.; Camphor Water to 1 oz. *Royal Chest.*
Chloride of Ammonium 20 grs.; Chloroform Water ½ oz.; Liquorice Water to 1 oz. *London.*

Mistura Anodyna.
Spirit of Camphor 30 mins.; Tincture of Henbane 30 mins.; Mucilage of Tragacanth 30 mins.; water to 1 oz. *Guy's.*

Mistura Antimonialis.
Tartar Emetic ⅙ gr.; Nitre 10 grs.; water to 1 oz. Dose ½ to 1 oz. *Consumption.*

Mistura Antimonii Aperiens.
Sulphate Magnesia 2 drms.; Tartarated Antimony $\frac{1}{12}$ gr.; water to 1 oz. *University.*

Mistura Aromatica.
Compound Powder of Cinnamon 7½ grs.; Spirit of Chloroform 20 mins.; water to 1 oz. For a dose. *Throat.*
Compound Powder of Cinnamon 7½ grs.; Pimento Water 1 oz. *London.*
Aromatic Powder of Chalk 15 grs.; Peppermint Water 1 oz. *St. Thomas's.*

Bicarbonate of Soda 10 grs. ; Carbonate of Ammonia
4 grs.; Compound Tincture of Cardamoms
½ drm.; Infusion of Cloves to 1 oz. *St. Mary's.*

Mistura Arsenicalis.

Arsenical Solution 4 mins. ; Burnt Sugar 10 mins. ;
Tincture of Ginger 10 mins.; water to 1 oz.
Charing Cross.
Arsenical Solution 5 mins.; water 1 oz. *British Skin.*
Arsenical Solution 5 mins.; Red Mixture to 1 oz. *London.*

Mistura Arsenici Composita.

Donovan's Solution 20 mins. ; Red Mixture to 1 oz. *London.*

Mistura Assafœtidæ.

Assafœtida 5 grs. ; water 1 oz. *British Skin.*
Tincture of Assafœtida 15 mins. ; Acacia Mixture to 1 oz. *London.*
Tincture of Assafœtida ½ drm. ; Tincture of Valerian 1 drm. ; Carbonate of Ammonia 4 grs.; Mucilage of Tragacanth ½ oz. ; water to 1 oz. *Charing Cross.*

Mistura Assafœtidæ c. Valerianâ.

Tincture of Assafœtida 30 mins. ; Ammoniated Tincture of Valerian ½ drm. ; Camphor Water to 1 oz. *London.*

Mistura Astringens.

Gallic Acid 3 grs. ; Diluted Sulphuric Acid 20 mins. ; Glycerine ½ drm. ; water ½ oz. *Samaritan.*
Aromatic Sulphuric Acid 15 mins. ; Spirit of Chloroform 20 mins. ; Compound Tincture of Camphor 60 mins. ; Decoction of Logwood to 1 oz. *London.*

Mistura Balsami Peruviani.

Peruvian Balsam 25 mins. ; Unclarified Honey 90 grs. ; water to 1 oz. *Guy's.*

Mistura Belladonnæ Composita.

Tincture of Belladonna 10 mins.; Dil. Hydrocyanic Acid 2 mins.; Bicarbonate of Soda 10 grs. ; Simple Syrup 60 mins. ; water to 1 oz. *Guy's.*

Mistura Benzoini Composita.
Comp. Tincture of Benzoin 20 mins.; Oxymel of Squill 30 mins.; Ipecacuanha Wine 5 mins.; Tincture of Tolu 5 mins.; water to 1 oz. *Consumption.*

Mistura Bismuthi.
White Bismuth 5 grs.; Carbonate of Magnesia 10 grs.; Mucilage of Tragacanth 30 mins.; Dill Water 1 oz. *King's.*
Carbonate of Bismuth 9 grs.; Mucilage of Tragacanth 1 drm.; water to 1 oz. *University.*
Subnitrate of Bismuth 10 grs.; Mucilage 1 drm.; Dill Water to 1 oz. *Royal Free.*
Carbonate of Bismuth 10 grs.; Mucilage 2 drms.; water 1 oz. *Samaritan.*
Nitrate of Bismuth 10 grs.; Bicarbonate of Soda 10 grs.; Powdered Calumba 10 grs.; Mucilage 1 drm.; Peppermint Water to 1 oz. *St. Thomas's.*
Subnitrate of Bismuth 10 grs.; Spirit of Chloroform 10 mins.; Tincture of Calumba 20 mins.; water to 1 oz. *Royal Chest.*
Subnitrate of Bismuth 10 grs.; Bicarbonate of Soda 10 grs.; Tragacanth 4 grs.; Cinnamon Water 1 oz. *Charing Cross.*
Subnitrate of Bismuth 10 grs.; Comp. Powder of Tragacanth 10 grs.; water to 1 oz. *Guy's.*
Subnitrate of Bismuth 10 grs.; Powder of Tragacanth 3 grs.; water 1 oz. *London Ophthalmic.*
White Bismuth 10 grs.; Carbonate of Magnesia 10 grs.; Mucilage 1 drm.; water to 1 oz. *Gt. Northern.*
Subnitrate of Bismuth 15 grs.; Comp. Tragacanth Powder 5 grs.; water to 1 oz. For a dose. *Throat.*
Subnitrate of Bismuth 15 grs.; Comp. Powder of Tragacanth 10 grs.; Peppermint Water to 1 oz. *London.*
Nitrate of Bismuth 15 grs.; Mucilage 1 drm.; Bicarbonate of Soda 10 grs.; Tincture of Calumba ½ drm.; water to 1 oz. *St. Mary's.*
Subnitrate of Bismuth 20 grs.; Mucilage of Tragacanth ½ oz.; Tincture of Cinnamon 10 mins.; water to 1 oz. *Consumption.*
Solution of Citrate of Bismuth and Ammonia 1 drm.; Diluted Hydrocyanic Acid 3 mins.; Peppermint Water 1 oz. *British Skin.*

Mistura Bismuthi Acida.
Nitrate of Bismuth (in crystals) 5 grs. ; Glycerine 1 drm.; water to 1 oz. *British Skin.*

Mistura Bismuthi Alkalina.
Subnitrate of Bismuth 10 grs.; Carbonate of Magnesia 15 grs.; Solution of Potash 10 mins.; Diluted Hydrocyanic Acid 4 mins.; Peppermint Water to 1 oz. *Throat.*

Mistura Bismuthi c. Acido Hydrocyanico.
Subnitrate of Bismuth 5 grs.; Dil. Hydrocyanic Acid 3 mins.; Comp. Infusion of Gentian 1 oz. *Consumption.*

Mistura Bismuthi Co.
Subnitrate of Bismuth 15 grs. : Carbonate of Magnesia 15 grs. ; Comp. powder of Tragacanth 10 grs. ; Tincture of Calumba 10 mins.; Chloroform water to 1 oz. *London.*

Mistura Bismuthi et Gentianæ.
Subnitrate of Bismuth 15 grs. ; Bicarbonate of Soda 15 grs.: Comp. Tragacanth powder 10 grs.; Comp. Infusion of Gentian to 1 oz. *London.*
Subnitrate of Bismuth 10 grs. ; Carbonate of Magnesia 5 grs.; Comp. Infusion of Gentian to 1 oz. *Consumption.*

Mistura Bismuthi Sedativa.
Subnitrate of Bismuth 10 grs.; Bicarbonate of Soda 10 grs.; Solution of Hydrochlorate of Morphia 10 mins.; Comp. Powd. of Tragac. 10 grs.; water to 1 oz. *Guy's.*

Mistura Buchu Co.
Bicarbonate of Potash 15 grs. ; Tincture of Hyoscyamus 20 mins.; Spirit of Chloroform 10 mins.; Infusion of Buchu to 1 oz. *London.*
Bicarbonate of Potash 10 grs. ; Tincture of Henbane 15 mins.; Infusion of Buchu 1½ oz. *St. Thomas's.*

Mistura Buchu c. Hyoscyamo.
Tincture of Henbane 12 mins. ; Mucilage 2 drms. Infusion of Buchu to 1 oz. *Royal Free.*

Mistura Calcis Hypophosphitis.
Hypophosphite of Lime 3 grs.; Glycerine 20 mins.; Tincture of Quassia 10 mins; water to ½ oz. *Consumption.*

Mistura Calcis et Sodæ Hypophosphitis.
Hypophosphite of Lime 5 grs.; Hypophosphite of Soda 5 grs.; Glycerine 15 mins.; water to 1 oz. *London.*

Mistura Calcis Sulphocarbolatis.
Sulpho-carbolate of Lime 10 grs.; Cassia water 1 oz. *London.*

Mistura Calumbæ Alkalina.
Bicarbonate of Soda 10 grs.; Infusion of Calumba to 1 oz. *Royal Chest.*

Mistura Calumbæ c. Sodâ.
Bicarbonate of Soda 15 grs.; Infusion of Calumba to 1 oz. *Royal Free.*

Mistura Camphoræ c. Æthere.
Spirit of Ether 30 mins.; Camphor Water to 1 oz. *London.*

Mistura Carminativa.
Carbonate of Magnesia 20 grs.; Powdered Rhubarb 10 grs.; Comp. Tincture of Camphor 30 mins.; Spirit of Sal Volatile 30 mins.; Dill Water to 1½ oz. *St. Thomas's.*
Carbonate of Magnesia 10 grs.; Comp. Tincture of Cardamoms 20 mins.; Tincture of Ginger 10 mins.; water to 1 oz. *Samaritan.*
Tincture of Ginger ½ drm.; Spirit of Chloroform 15 mins.; Essence of Cloves 6 mins.; Tincture of Opium 5 mins.; Peppermint Water to 1 oz. *Royal Free.*

Mistura Caryophylli Composita.
Bicarbonate of Soda 10 grs.; Aromatic Spirit of Ammonia 20 mins.; Spirit of Chloroform 15 mins.; Infusion of Cloves to 1 oz. *St. George's.*

Mistura Cascarillæ Acida.
Dil. Nitric Acid 15 mins.; Tincture of Hops 30 mins.; Infusion of Cascarilla to 1 oz. *University.*

Dil. Nitric Acid 15 mins.; Comp. Mixture of Cascarilla to 1 oz. *London.*

Mistura Cascarillæ c. Acido Nitrico.
Dil. Nitric Acid 15 mins.; Tincture of Hops 30 mins.; Infusion of Cascarilla to 1 oz. *Consumption.*

Mistura Cascarillæ c. Ammoniâ.
Carbonate of Ammonia 3 grs.; Compound Spirit of Horseradish 10 mins.; Infusion of Cascarilla to 1 oz. *Consumption.*

Mistura Cascarillæ Co.
Vinegar of Squill 10 mins.; Comp. Tincture of Camphor 15 mins.; Infusion of Cascarilla to 1 oz. *St. Thomas's.*
Tincture of Squill 15 mins.; Comp. Tincture of Camphor 60 mins.; Infusion of Cascarilla to 1 oz. *Guy's.*
Tincture of Squill 15 mins.; Compound Tincture of Camphor 15 mins.; Infusion of Cascarilla to 1 oz. *Royal Free.*
Vinegar of Squill 15 mins.; Comp. Tincture of Camphor 30 mins.; Infusion of Cascarilla to 1 oz. *London.*
Vinegar of Squill 30 min.; Comp. Tincture of Camphor 60 mins.; Infusion of Cascarilla to 12 drms. *St. George's.*

Mistura Cascarillæ c. Scillâ.
Oxymel of Squill ½ drm.; Infusion of Cascarilla to 1 oz. *Royal Chest.*

Mistura Catechu.
Tincture of Catechu 1 drm.; Bicarbonate of Potash 10 grs.; Cinnamon Water to 1 oz. *British Skin.*

Mistura Catechu et Cretæ.
Tincture of Catechu 30 mins.; Chalk Mixture to 1 oz. *University.*

Mistura Cerulea.
Tincture of Litmus 5 mins.; Water 1 oz. *Royal Free.*

Mistura Cetacei.
Spermaceti 15 grs.; Chloric Ether 5 mins.; Comp. Tragacanth Powder 20 grs.; Pimento Water to 1 oz. *Consumption.*
Spermaceti 20 grs.; Powdered Gum Acacia 20 grs.; water to 1 oz. *Royal Chest.*

Mistura Cetacei Opiata.
Spermaceti 15 grs.; Ipecacuanha Wine 10 mins.; Comp. Tincture of Camphor 30 mins.; Comp. Tragacanth Powder 20 grs.; water to 1 oz. *Consumption.*

Mistura Chalybeata.
Tincture of Perchloride of Iron 10 mins.; Dil. Nitric Acid 15 mins.; Syrup 20 mins.; Infusion of Quassia to 1 oz. *St. Thomas's.*

Mistura Chloral.
Hydrate of Chloral 10 grs.; Spirit of Lemon 3 mins.; Chloroform water ½ oz.; water to 1 oz. *London.*

Mistura Chloral. Composita.
Chloral Hydrate 10 grs.; Bromide of Potassium 10 grs.; Syrup of Orange 30 mins.; water to 1 oz. *Consumption.*

Mistura Chloralis Hydratis.
Hydrate of Chloral 10 grs.; Syrup of Red Poppies 1 drm.; water to 1 oz. *British Skin.*
Hydrate of Chloral 20 grs.; Syrup of Orange Flower 40 mins.; Syrup of Tolu 40 mins.; water to 1 oz. *Throat.*

Mistura Chlori.
Solution of Chlorine 10 mins.; water to 1 oz. *London.*

Mistura Chloroformi.
Chloroform Water ½ oz.; water to 1 oz. *British Skin.*
Spirit of Chloroform 10 mins.; Mucilage of Tragacanth 1 oz. *King's.*

Mistura Chloroformi c. Ammoniâ.
Spirit of Chloroform 15 mins.; Carbonate of Ammonia 3 grs.; Infusion of Gentian to 1 oz. Dose 1 oz. *London Ophthalmic.*

Mistura Cinchonæ Acida.
Diluted Sulphuric Acid 10 mins. ; Comp. Tincture of Gentian ½ drm.; Decoction of Cinchona to 1 oz. *Royal Free.*
Diluted Nitro-hydrochloric Acid 10 mins.; Decoction of Cinchona to 1 oz. *London.*
Diluted Nitric Acid 10 mins. ; Decoction of Yellow Cinchona ½ oz.; Water to 1 oz. *Throat.*

Mistura Cinchonæ c. Acido.
Dil. Sulphuric Acid 10 mins. ; Decoction of Bark 1 oz. *St. Thomas's.*

Mistura Cinchonæ c. Acid. Nitric.
Dil. Nitric Acid 15 mins. ; Decoction of Yellow Bark to 1 oz. *Consumption.*

Mistura Cinchonæ c. Acid. Sulph.
Dil. Sulphuric Acid 15 mins. ; Decoction of Yellow Bark to 1 oz. *Consumption.*

Mistura Cinchonæ Ammoniata.
Bicarbonate of Soda 10 grs. ; Aromatic Spirit of Ammonia 20 mins.; Decoction of Yellow Cinchona ½ oz. ; water to 1 oz. *Throat.*

Mistura Cinchonæ c. Ammoniâ.
Carbonate of Ammonia 5 grs. ; Decoction of Yellow Bark to 1 oz. *Consumption.*

Mistura Cinchonæ Co.
Carbonate of Magnesia 7 grs. ; Sulphate of Magnesia 20 grs. ; Decoction of Yellow Bark 1 oz. *St. George's.*
Carbonate of Ammonia 4 grs.; Decoction of Cinchona Bark 1 oz. *Charing Cross.*

Mistura Cinchoniæ.
Hydrochlorate of Cinchonia 2 grs. ; Dil. Hydrochloric Acid 2 mins. ; water 1 oz. *University.*
Sulphate of Cinchonia 2 grs.; Dil. Sulphuric Acid 5 mins.; water to 1 oz. *Royal Chest.*
Dil. Sulphuric Acid 10 mins.; Decoction of Cinchona 1 oz. *Gt. Northern.*

Mistura Cinchoniæ Acida.
Cinchonine 1 gr.; Dil. Sulphuric Acid 5 mins.; water 1 oz. *St. Mary's.*

Mistura Cinchoninæ.
Sulphate of Cinchonine 1 gr.; Dil. Sulphuric Acid 5 mins.; water to 1 oz. *London.*
Hydrochlorate of Cinchonine 2 grs.; Dil. Hydrochloric Acid 4 mins.; Tincture of Orange 10 mins.; Simple Syrup 30 mins.; water to 1 oz. *Guy's.*

Mistura Cinchonidinæ Sulphatis.
Sulphate of Cinchonidine 1 gr.; Dil. Sulphuric Acid 5 mins.; water to 1 oz. *Royal Free.*

Mistura Cinchonidinæ c. Ferro.
Sulphate of Cinchonidine 1 gr.; Sulphate of Iron 1 gr.; Dil. Sulphuric Acid 3 mins.; water to 1 oz. *Royal Free.*

Mistura Colchici.
Wine of Colchicum 15 mins; Carbonate of Magnesia 10 grs.; Pimento Water 1 oz. *Westminster Ophthalmic.*
Wine of Colchicum 15 mins.; Carbonate of Magnesia 10 grs.; Spirit of Nitrous Ether 30 mins.; water to 1 oz. *Royal Free.*
Wine of Colchicum 15 mins.; Carbonate of Magnesia 10 grs.; Bicarbonate of Potash 10 grs.; Peppermint Water 1 oz. *St. Thomas's.*
Wine of Colchicum 20 mins.; Carbonate of Magnesia 10 grs.; Pimento Water to 1 oz. Dose 1 oz. *London Ophthalmic.*
Tincture of Colchicum Seeds 15 mins.; Bicarbonate of Potash 10 grs.; Pimento Water to 1 oz. *London.*

Mistura Colchici Alkalina.
Tincture of Colchicum Seeds 6 mins.; Bicarbonate of Potash 5 grs.; Pimento Water 1 oz. *British Skin.*
Tincture of Colchicum Seeds 24 mins.; Bicarbonate of Soda 7½ grs.; water to 1 oz. *University.*

Mistura Colchici Aperiens.
Tincture of Colchicum Seeds 15 mins.; Carbonate of Magnesia 6 grs.; Sulphate of Magnesia 30 grs.; Peppermint Water to 1 oz. *University.*
Sulphate of Magnesia 60 grs.; Colchicum Mixture to 1 oz. *London.*

Mistura Colchici et Magnesiæ.
Wine of Colchicum 15 mins.; Carbonate of Magnesia 12½ grs.; Sulphate of Magnesia 20 grs.; Infusion of Cloves 1 oz. *St. George's.*

Mistura Colchici et Potassæ.
Wine of Colchicum 15 mins.; Solution of Potash 30 mins.; Pimento Water to 1 oz. *St. George's.*

Mistura Conii Co.
Extract of Conium 5 grs.; Bicarbonate of Soda 5 grs.; Spirit of Nutmeg 30 mins.; Liquorice Water to 1 oz. *Guy's.*

Mistura Conii c. Hyoscyamo.
Juice of Conium 30 mins.; Extract of Henbane 3 grs.; Mucilage of Acacia 2 drms.; water to 1 oz. *Royal Chest.*

Mistura Copaibæ.
Copaiba 15 mins.; Mucilage 2 drms.; water to 1 oz. *St. Thomas's.*
Copaiba 20 mins.; Tincture of Quillaia 20 mins.; Spirit of Nitrous Ether 30 mins.; Camphor Water to 1 oz. *London.*
Copaiba 20 mins.; Mucilage 2 drms.; Dil. Sulphuric Acid 10 mins.; water to 1 oz. *Royal Free.*
Copaiba 20 mins.; Solution of Potash 20 mins.; Tincture of Opium 5 mins.; Peppermint Water 1 oz. *Gt. Northern.*
Copaiba 20 mins.; Mucilage 1 drm.; Chloric Ether 5 mins.; Camphor Water to 1 oz. *Consumption.*
Copaiba 20 mins.; Tincture of Quillaia 20 mins.; Tincture of Lemon Peel 20 mins.; water to 1 oz. *Guy's.*
Copaiba 20 mins.; Solution of Potash 10 mins.; Pimento Water to 1 oz. *London Ophthalmic.*
Copaiba 20 mins.; Dil. Sulphuric Acid 10 mins.; Mucilage ½ oz.; water to 1 oz. *King's.*
Copaiba 24 mins.; Mucilage 48 mins.; Dil. Sulphuric Acid 12 mins.; Acid Infusion of Roses 1 oz. *St. George's.*
Copaiba 30 mins.; Mucilage 1 drm.; Pimento Water to 1 oz. *St. Mary's.*
Copaiba 30 mins.; Solution of Potash 12 mins.; Cinnamon Water to 1 oz. *University.*

Mistura Copaibæ Co.
    Oil of Cubebs 10 mins.; Copaiba Mixture to 1 oz.
    London.

Mistura Copaibæ Resinæ.
    Resin of Copaiva 15 grs.; Rectified Spirit of Wine
    20 mins.; Comp. Powder of Tragacanth 15 grs.;
    Syrup of Ginger 60 mins.; water to 1 oz.
    Guy's.
    Resin of Copaiva 15 grs.; Almond Oil 40 mins.;
    Tincture of Quillaia 10 mins.; Syrup of Ginger
    30 mins.; water to 1 oz. Mix the Resin with
    the oil on a water-bath, then triturate them
    with the tincture, then add the water and
    syrup gradually. *London*.

Mistura Creasoti.
    Creasote 1 min.; Comp. Powder of Tragacanth 5
    grs.; Spirit of Juniper 20 mins.; Chloroform
    Water ½ oz.; water to 1 oz. *London*.

Mistura Creasoti c. Opio.
    Creasote 1 min.; Tincture of Opium 2 mins.; Spirit
    of Chloroform 15 mins.; Glycerine 1 drm.;
    water to 1 oz. *Royal Chest*.
    Creasote 1 min.; Tincture of Opium 3 mins.; Syrup
    of Orange 30 mins.; Mucilage 1 drm.; Cinnamon Water to 1 oz. *Consumption*.

Mistura Cretæ.
    Prepared Chalk 15 grs.; Gum Acacia 15 grs.; Cassia
    Water to 1 oz. *London*.
    Prepared Chalk 15 grs.; Gum Acacia 15 grs.;
    Cinnamon Water 1 oz. *Throat*.

Mistura Cretæ Aromatica.
    Comp. Powder of Cinnamon 10 grs.; Powdered
    Acacia 10 grs.; Prepared Chalk 10 grs.; water
    1 oz.; *London Ophthalmic*.
    Comp. Powder of Cinnamon 7½ grs.; Chalk Mixture (Hosp.) 1 oz. *London*.

Mistura Cretæ Aromat. Anodyna.
    Comp. Powder of Cinnamon 10 grs.; Powdered
    Acacia 10 grs.; Prepared Chalk 10 grs.; Tincture of Opium 5 mins.; water 1 oz. *London Ophthalmic*.

Mistura Cretæ c. Catechu.
Tincture of Catechu 30 mins. ; Chalk Mixture to 1 oz. *Consumption.*
Tincture of Catechu 1 drm. ; Spirit of Chloroform 10 mins. ; Chalk Mixture to 1 oz. *St. Mary's.*

Mistura Cretæ Composita.
Aromatic Spirit of Ammonia 20 mins. ; Tincture of Catechu 20 mins. ; Mixture of Chalk to 1 oz. *Guy's.*

Mistura Cretæ c. Hæmatoxylo.
Chalk Mixture ½ oz. ; Tincture of Catechu 30 mins. ; Decoction of Logwood to 1 oz. for an adult. *Gt. Northern. King's.*
Extract of Logwood 10 grs. ; Chalk Mixture (Hosp.) 1 oz. *London.*

Mistura Cretæ Opiata.
Aromatic Chalk Powder 20 grs. ; Tincture of Opium 10 mins. ; Chalk Mixture 1 oz. *St. Mary's.*

Mistura Cretæ Co. c. Opio.
Aromatic Powder of Chalk 20 grs. ; Tincture of Opium 5 mins. ; Spirit of Chloroform 20 mins. ; Tincture of Capsicum 2 mins. ; Pimento Water to 1 oz. *Royal Chest.*

Mistura Cretæ c. Opio.
Tincture of Catechu 30 mins. ; Tincture of Opium 5 mins. ; Chalk Mixture to 1 oz. *Consumption.*
Comp. Chalk Powder P. L. 20 grs. ; Tincture of Opium 5 mins. ; water 1 oz. *St. Thomas's.*
Tincture of Opium 5 mins. ; Chalk Mixture (Hosp.) to 1 oz. *London.*
Tincture of Opium 10 mins. ; Tincture of Rhatany 1 drm. ; Chalk Mixture to 1 oz. *Charing Cross.*

Mistura Croton-Chloral.
Croton-Chloral Hydrate 4 grs. ; Glycerine 15 mins. ; water to 1 oz. *Throat.*
Croton-Chloral Hydrate 4 grs. ; Glycerine 15 mins. : Chloroform water ½ oz. ; water to 1 oz. *London.*

Mistura Diaphoretica.
Ipecacuanha Wine 5 mins. ; Spirit of Nitrous Ether 30 mins. ; Solution of Acetate of Ammonia 2 drms. ; water to 1 oz. *British Skin.*

Ipecacuanha Wine 8 mins.; Tincture of Opium 5 mins.; Bicarbonate of Potash 20 grs.; Water to 1 oz. *Royal Free.*

Ipecacuanha Wine 15 mins.; Spirit of Nitrous Ether 30 mins.; Saline Mixture to 1 oz. *London.*

Vinegar of Ipecacuanha 15 mins.; Spirit of Nitrous Ether 30 mins.; Saline Mixture to 1 oz. *Throat.*

Mistura Diuretica.

Acetate of Potash 20 grs.; Vinegar of Squill 20 mins.; Decoction of Broom to 1 oz. *London. Throat.*

Acetate of Potash 20 grs.; Spirit of Nitrous Ether 30 mins.; Spirit of Juniper 30 mins.; Mucilage 1 drm.; Decoction of Broom to 1 oz. *St. Mary's.*

Acetate of Potash 20 grs.; Spirit of Nitrous Ether 30 mins.; Decoction of Broom to 1 oz. *Charing Cross.*

Acetate of Potash 20 grs.; Vinegar of Squill 20 mins.; Tincture of Digitalis 5 mins.; water to 1 oz. *British Skin.*

Acetate of Potash 30 grs.; Tincture of Squill 15 mins.; Spirit of Nitrous Ether 30 mins.; Spirit of Juniper 30 mins.; Decoction of Broom to 1 oz. *Royal Free.*

Acetate of Potash 30 grs.; Tincture of Squill 20 mins.; Sweet Spirit of Nitre 30 mins.; Decoction of Broom to 1 oz. *Gt. Northern.*

Nitrate of Potash 6 grs.; Bicarbonate of Potash 10 grs.; Carbonate of Ammonia 3 grs.; Spirit of Juniper 1 drm.; water to ½ oz. *Samaritan.*

Mistura Dyspeptica.

Bicarbonate of Soda 10 grs.; Sulphate of Magnesia 30 grs.; Essence of Peppermint 10 mins.; Comp. Infusion of Gentian 1 oz. *St. Thomas's.*

Mistura Effervescens.

Bicarbonate of Potash 15 grs.; water 1 oz. (Citric Acid 10 grs.; water 1 oz.). *Gt. Northern.*

Bicarbonate of Potash 20 grs.; water to 1 oz. during effervescence with 15 grs. of Citric Acid. *Throat.*

Bicarbonate of Soda 20 grs.; Water 1 oz. with ½ oz. Citric Acid Solution in effervescence. *London.*

Bicarbonate of Soda 20 grs. ; water 1 oz. (Tartaric Acid 18 grs. ; water ½ oz.). *King's. Royal Free.*

Mistura Emolliens.

Nitrate of Potash 10 grs. ; Sulphate of Magnesia ½ drm. ; Tincture of Henbane 15 mins. ; Mucilage 2 drms. ; water to 1½ oz. *St. Thomas's.*

Mistura Ergotæ.

Fluid Extract of Ergot 30 mins. ; Glycerine 20 mins. ; Tincture of Cinnamon 15 mins. ; water to 1 oz. *Consumption.*

Fluid Extract of Ergot 20 mins. ; Aromatic Sulphuric Acid 10 mins. ; Cinnamon Water 1 oz. *Samaritan.*

Fluid Extract of Ergot 30 mins. ; Glycerine 20 mins. ; Cinnamon Water to 1 oz. *Charing Cross.*

Fluid Extract of Ergot 20 mins. ; Sal Volatile 10 mins. ; water 1 oz. *Gt. Northern.*

Liquid Extract of Ergot 30 mins.; Tincture of Orange Peel 30 mins. ; water to 1 oz. *Guy's.*

Mistura Ergotæ Ammoniata.

Ammoniated Extract of Ergot 20 mins. ; Spirit of Chloroform 15 mins. ; Camphor Water to 1 oz. *University.*

Ammoniated Extract of Ergot 30 mins. ; Cassia Water to 1 oz. *London.*

Mistura Ergotæ Co.

Liquid Extract of Ergot 15 mins. ; Sulphate of Magnesia 1 drm. ; Gallic Acid 10 grs. ; Dil. Sulphuric Acid 5 mins. ; water to 1 oz. *Royal Chest.*

Liquid Extract of Ergot 40 mins. ; Gallic Acid 10 grs. ; Cassia Water to 1 oz. *London.*

Mistura Ergotæ c. Ferro.

Tincture of Perchloride of Iron 15 mins. ; Liquid Extract of Ergot 15 mins. ; Spirit of Chloroform 15 mins. ; Infusion of Quassia 1 oz. *St. Mary's.*

Solution of Perchloride of Iron 15 mins. ; Liquid Extract of Ergot 20 mins.; Spirit of Chloroform 10 mins. ; Infusion of Quassia to 1 oz. *London.*

Mistura Ergotæ Plumbea.
Acetate of Lead 2 grs. ; Liquid Extract of Ergot 16 mins. ; Distilled Water to 1 oz. *Throat.*

Mistura Expectorans.
Ipecacuanha Wine 10 mins. ; Tincture of Squill 20 mins. ; Spirit of Nitrous Ether 30 mins. ; Chloroform Water 2 drms. ; water to 1 oz. *Charing Cross.*

Ipecacuanha Wine 15 mins. ; Nitrate of Potash 6 grs. ; Compound Tincture of Camphor 30 mins. ; Mucilage 1 drm. ; water to 1 oz. *Royal Free.*

Ipecacuanha Wine 30 mins. ; Antimonial Wine 30 mins. ; Spirit of Chloroform 30 mins. ; Tincture of Belladonna 15 mins. ; Syrup of Squill 3 drms. ; Mindererus Spirit to 1 oz. Dose 1 to 2 drms. *Samaritan.*

Tincture of Squill 20 mins. ; Paregoric 30 mins. ; Infusion of Cascarilla to 1 oz. *Gt. Northern.*

Mistura Febrifuga.
Nitrate of Potash 7 grs. ; Spirit of Nitrous Ether 30 mins. ; Solution of Acetate of Ammonia ½ oz. ; water to 1 oz. *Royal Free.*

Mistura Ferri.
Tincture of Perchloride of Iron 10 mins. ; Glycerine ½ drm. ; water to 1 oz. *Royal Chest.*

Mistura Ferri Acida.
Sulphate of Iron 2 grs. ; Sulphate of Magnesia 20 grs. ; Diluted Sulphuric Acid 5 mins. ; Dill Water ½ oz. *Samaritan.*

Solution of Perchloride of Iron 15 mins. ; Diluted Nitro-Hydrochloric Acid 10 mins. ; water to 1 oz. *Consumption.*

Mistura Ferri c. Æther. Chlor.
Solution of Perchloride of Iron 20 mins. ; Chloric Ether 10 mins. ; Infusion of Quassia to 1 oz. Dose ½ to 1 oz. *Consumption.*

Mistura Ferri c. Aloe.
Comp. Mixture of Iron P. B. 6 drms. ; Comp. Decoction of Aloes to 1 oz. *Consumption.*

Comp. Iron Mixture 4 drms. ; Comp. Decoction of Aloes to 1 oz. *London.*

Mistura Ferri Amara.
Solution of Perchloride of Iron 30 mins.; Spirit of Chloroform 5 mins.; Infusion of Quassia to 1 oz. *University.*

Mistura Ferri Ammon. Citrat.
Ammonio-Citrate of Iron 5 grs.; Pimento Water 1 oz. *London.*
Ammonio-Citrate of Iron 5 grs.; Camphor Water 1 oz. *Throat.*
Ammonio-Citrate of Iron 5 grs.; Water to 1 oz. *Royal Free.*

Mistura Ferri Ammonio-Tartratis.
Ammonio-Tartrate of Iron 5 grs.; water 1 oz. *Westminster Ophthalmic.*

Mistura Ferri Aperiens.
Sulphate of Iron 1 gr.; Sulphate of Magnesia 30 grs.; Dil. Sulphuric Acid 5 mins.; water 1 oz. *London Ophthalmic.*
Sulphate of Iron 1½ gr.; Sulphate of Magnesia 30 grs.; Dil Sulphuric Acid 8 mins.; Spirit of Chloroform 8 mins.; Peppermint Water 1 oz. *Westminster Ophthalmic.*
Sulphate of Iron 2 grs.; Sulphate of Magnesia 30 grs.; Dil. Sulphuric Acid 9 mins.; Peppermint Water to 1 oz. *University.*
Sulphate of Iron 2 grs.; Sulphate of Magnesia 30 grs.; Dil. Sulphuric Acid 5 mins.; Peppermint Water to 1 oz. *Throat.*

Mistura Ferri Arsenicalis.
Ammonio-Citrate of Iron 7½ grs.; Arsenical Solution 5 mins.; Tincture of Calumba 24 mins.; water to 1 oz. *University.*
Solution of Arsenic 5 mins.; Citrate of Iron and Cinchonine 5 grs.; Distilled Water to 1 oz. *London.*

Mistura Ferri et Bismuthi.
Citrate of Iron and Ammonia 5 grs.; Subnitrate of Bismuth 5 grs.; water to 1 oz. *Royal Chest.*

Mistura Ferri c. Calumbâ.
Ammonio-Citrate of Iron 5 grs.; Stronger Solution of Ammonia 1½ min.; Spirit of Nutmeg 6 mins.; Infusion of Calumba to 1 oz. *Consumption.*

Mistura Ferri Cathartica.
Sulphate of Iron 3 grs.; Sulphate of Magnesia 60 grs.; Dil Sulphuric Acid 10 mins.; Peppermint Water to 1 oz. *London.*

Mistura Ferri Citratis c. Ammoniâ.
Ammonio-Citrate of Iron 5 grs.; Aromatic Spirit of Ammonia 15 mins.; Chloroform Water 2 drms.; Infusion of Chiretta to 1 oz. *Charing Cross.*

Mistura Ferri Co.
Sulphate of Iron 2 grs.: Myrrh 5 grs.; Carbonate of Potash 2½ grs.; Sugar 5 grs.; Essence of Cassia 5 mins.; water to 1 oz. *London.*
Sulphate of Iron 2½ grs.; Myrrh 6 grs.; Carbonate of Potash 3 grs.; Sugar 3 grs.; Spirit of Nutmeg 24 mins.; water to 1 oz. *London Ophthalmic.*
Sulphate of Iron 2½ grs.; Myrrh 5 grs.; Carbonate of Potash 3 grs.; White Sugar 6 grs.; Cinnamon Water 1 oz. *Charing Cross.*
Sulphate of Iron 4 grs.; Sugar 10 grs.; water 2 drms.: dissolve, and add Carbonate of Potash 5 grs.; Spirit of Nutmeg 30 mins.; water 2 drms. *Samaritan.*

Mistura Ferri et Cinchoninæ Citratis.
Citrate of Iron and Cinchonine 5 grs.; water 1 oz. *London.*

Mistura Ferri Dialysati.
Dialysed Iron 10 mins.; Distilled Water to 1 oz. *London.*

Mistura Ferri c. Hydrargyro.
Solution of Perchloride of Iron 10 mins.; Solution of Perchloride of Mercury P. B. ½ to 1 drm.; Infusion of Quassia to 1 oz. *Samaritan*

Mistura Ferri Laxans.
Sulphate of Iron 2 grs.; Sulphate of Magnesia 60 grs.; Sulphuric Acid 3 mins.; Peppermint Water to 1 oz. *St. Mary's.*

Mistura Ferri c. Magnesiæ Sulphate.
Sulphate of Iron 2 grs.; Sulphate of Magnesia 20 grs.; water 1 oz. *King's.*

K

Sulphate of Iron 2 grs.; Sulphate of Magnesia 30 grs.; Dil. Sulphuric Acid 10 mins.; Peppermint Water to 1 oz. *Charing Cross.*
Sulphate of Iron 2 grs.; Sulphate of Magnesia 30 grs.; Dil. Sulphuric Acid 10 mins.; Tincture of Quassia 20 mins.; water to 1 oz. *Consumption.*

Mistura Ferri c. Nuce Vomicâ.

Tincture of Nux Vomica 10 mins.; Ammonio-Citrate of Iron 5 grs.; Carbonate of Ammonia 4 grs.; Camphor Water 1 oz. *Samaritan.*

Mistura Ferri Opiata.

Solution of Perchloride of Iron 20 mins.; Tincture of Opium 1½ min.; Chloroform Water 2 drms.; water to 1 oz. *Throat.*

Mistura Ferri c. Opio.

Ammonio-Citrate of Iron 5 grs.; Tincture of Opium 4 mins.; water 1 oz. *British Skin.*

Mistura Ferri Perchloridi.

Solution of Perchloride of Iron 15 mins.; Infusion of Quassia 1 oz. *Royal Free.*
Solution of Perchloride of Iron 15 mins.; Glycerine 18 mins.; Spirit of Chloroform 9 mins.; water to 1 oz. *University.*
Solution of Perchloride of Iron 15 mins.; Spirit of Chloroform 15 mins.; water 1 oz. *St. Mary's.*
Solution of Perchloride of Iron 15 mins.; Red Mixture to 1 oz. *London.*
Solution of Perchloride of Iron 15 mins.; Sulphate of Magnesia 15 grs.; water 1 oz. *British Skin.*
Solution of Perchloride of Iron 15 mins.; Chloroform Water 3 drms.; Infusion of Quassia to 1 oz. *Charing Cross.*
Solution of Perchloride of Iron 10 mins.; Infusion of Quassia 1 oz. *King's.*
Solution of Perchloride of Iron 10 mins.; Glycerin 10 mins.; Infusion of Quassia ½ oz.; water ½ oz. *Westminster Ophthalmic.*
Tincture of Perchloride of Iron 10 mins.; Glycerine 15 mins.; water to 1 oz. Dose 1 oz. *London Ophthalmic.*

Mistura Ferri c. Potassæ Chlorate.
Chlorate of Potash 5 grs.; Tincture of Perchloride of Iron 15 mins.; Chloric Ether 5 mins.; water to 1 oz. *Consumption.*

Mistura Ferri Perchlor. c. Potassæ Chlorate.
Tinct. of Perchloride of Iron 10 mins.; Chlorate of Potash 5 grs.; water 1 oz. *Samaritan.*

Mistura Ferri Perchlor. c. Quassiâ.
Solution of Perchloride of Iron 20 mins.; Chloroform Water 2 drms.; Infusion of Quassia to 1 oz. Dose 1 oz. *Throat.*

Mistura Ferri Perchlor. c. Quiniâ.
Solution of Perchloride of Iron 10 mins.; Sulphate of Quinia 1 gr.; Cinnamon Water 1 oz.; *Samaritan.*

Mistura Ferri c. Potassii Iodido.
Tartarated Iron 5 grs.; Iodide of Potassium 2½ grs.; Pimento Water to 1 oz. *London.*
Tartarated Iron 20 grs.; Iodide of Potassium 10 grs.; water 1 oz. *King's.*
Ammonio-Citrate of Iron 10 grs.; Aromatic Spirit of Ammonia 20 mins.; Iodide of Potassium 5 grs.; water to 1 oz. *University.*

Mistura Ferri c. Quassiâ.
Solution of Perchloride of Iron 15 mins.; Spirit of Cinnamon 20 mins.; Infusion of Quassia to 1 oz. *Consumption.*
Tincture of Perchloride of Iron 15 mins.; Infusion of Quassia to 1 oz. *Guy's.*
Ammonio-Citrate of Iron 10 grs.; Infusion of Quassia 1 oz. *Charing Cross.*

Mistura Ferri c. Quiniâ.
Sulphate of Iron 2 grs.; Sulphate of Quinine 2 grs.; Dil. Sulphuric Acid 10 mins.; water ½ oz. *Samaritan.*

Mistura Ferri et Quiniæ Effervescens.
Ammonio-Citrate of Iron 5 grs.; Sulphate of Quinia 1 gr.; Citric Acid 10 grs.; water 1 oz. To be taken with 10 grs. of Bicarbonate of Soda. *Consumption.*

Mistura Ferri Salina.

Solution of Perchloride of Iron 24 mins.; Citrate of Potash 22 grs.; Spirit of Chloroform 9 mins.; water to 1 oz. *University*.

Ammonio-Citrate of Iron 5 grs.; Citrate of Potash 20 grs.; Chloroform Water to 1 oz. *London*.

Mistura Ferri c. Strychniâ.

Solution of Perchloride of Iron 10 mins.; Solution of Strychnia 4 mins.; Spirit of Chloroform 10 mins.; water 1 oz. *Westminster Ophthalmic*.

Solution of Strychnia 5 mins.; Mixture of Perchloride of Iron to 1 oz. *London*.

Mistura Ferri Sulph.

Sulphate of Iron 1 gr.; Dil. Sulphuric Acid 5 mins.; Infusion of Quassia 1 oz. *Gt. Northern*.

Sulphate of Iron 2 grs.; Dil. Sulphuric Acid 5 mins.; water to 1 oz. *London*.

Sulphate of Iron 2 grs.; Dil. Sulphuric Acid 5 mins.; Infusion of Quassia to 1 oz. *Royal Free*.

Sulphate of Iron 3 grs.; Dil. Sulphuric Acid 15 mins.; water to 1 oz. *Royal Chest*.

Mistura Ferri Tartarati.

Tartarated Iron 10 grs.; Rectified Spirit of Wine 30 mins.; Simple Syrup 60 mins.; water to 1 oz. *Guy's*.

Mistura Ferri et Valerianæ.

Tincture of Valerian 20 mins.; Ammonio-Citrate of Iron 5 grs.; Pimento Water to 1 oz. *London*

Mistura Filicis vel Haustus Filicis.

Liquid Extract of Male Fern 30 mins.; Powdered Tragacanth 10 grs.; Peppermint Water to 1 oz. *London Ophthalmic*.

Liquid Extract of Male Fern 60 mins.; Syrup of Ginger 30 mins.; Tincture of Quillaia 30 mins.; water to 1 oz. *Guy's*.

Oil of Male Fern 2 drms.; Comp. Tragacanth Powder 1 drm.; Peppermint Water to 2 oz. Dose 2 oz. *St. Mary's*.

Mistura Flava.

- Caramel *q. s.*; water 1 oz. *University*.

Tincture of Turmeric 5 mins.; water to 1 oz. *Royal Free*.

Mistura Gelsemini.
Tincture of Yellow Jessamine 15 mins.; Infusion of Cascarilla to 1 oz. *Royal Chest.*
Tincture of Gelsemium 10 mins.; Red Mixture to 1 oz. *London.*

Mistura Gentianæ Acida.
Dil. Hydrochloric Acid 10 mins.; Comp. Infusion of Gentian to 1 oz. *Consumption.*
Dil. Sulphuric Acid 10 mins.; Comp. Infusion of Gentian to 1 oz. *London.*
Diluted Nitro-Hydrochloric Acid 12 mins.; Spirit of Chloroform 10 mins.; Comp. Infusion of Gentian 1 oz. *University.*

Mistura Gentianæ c. Acido Nitrico.
Diluted Nitric Acid 10 mins.; Comp. Infusion of Gentian 1 oz. *Gt. Northern.*

Mistura Gentianæ Alkalina.
Dil. Hydrocyanic Acid 3 mins.; Bicarbonate of Soda 15 grs.; Comp. Infusion of Gentian to 1 oz. *Consumption.*
Bicarbonate of Soda 15 grs.; Infusion of Gentian 1 oz. *Charing Cross.*

Mistura Gentianæ Cathartica.
Sulphate of Magnesia 60 grs.; Comp. Infusion of Gentian 1 oz. *London. London Ophthalmic.*

Mistura Gentianæ Composita.
Comp. Tincture of Cardamoms 80 mins.; Comp. Infusion of Gentian 4 drms.; Infusion of Senna 4 drms. *Guy's.*

Mistura Gentianæ c. Ammoniâ.
Carbonate of Ammonia 3 grs.; Comp. Infusion of Gentian 1 oz. *Royal Chest.*
Carbonate of Ammonia 4 grs.; Bicarbonate of Soda 10 grs.; Comp. Infusion of Gentian 1 oz. *London Ophthalmic.*
Carbonate of Ammonia 5 grs.; Comp. Infusion of Gentian 1 oz. *Gt. Northern.*
Aromatic Spirit of Ammonia 20 mins.; Bicarbonate of Soda 10 grs.; Comp. Infusion of Gentian to 1 oz. *Consumption.*

Mistura Gentianæ et Magnesiæ.
Carbonate of Magnesia 15 grs.; Sulphate of Magnesia 60 grs.; Tincture of Gentian 20 mins.; Infusion of Gentian to 1 oz. *St. George's.*

Mistura Gentianæ c. Rheo.
Gentian Root 90 grs.; Bitter Orange Peel 30 grs.; Rhubarb Root 20 grs.; Ginger Root 15 grs.; Boiling Water 20 oz. Infuse 2 hours and strain. Dose 1 to 2 oz. *London.*

Mistura Gentianæ et Sennæ.
Infusion of Gentian 6 drms.; Infusion of Senna 3 drms.; Comp. Tincture of Cardamoms 1 drm. *St. George's.*

Mistura Gentianæ c. Sodâ.
Bicarbonate of Soda 10 grs.; Spirit of Chloroform 15 mins.; Comp. Infusion of Gentian to 1 oz. *Throat.*
Comp. Infusion of Gentian 1 oz.; Bicarbonate of Soda 18 grs. *University.*
Comp. Infusion of Gentian 1 oz.; Bicarbonate of Soda 10 grs. *London.*

Mistura Guaiaci.
Ammoniated Tincture of Guaiacum 30 mins.; Acacia Mixture to 1 oz. *London.*

Mistura Guaiaci Ammoniata.
Ammoniated Tincture of Guaiacum 30 mins.; Mucilage of Acacia 30 mins.; water 1 oz. *Throat.*

Mistura Guaiaci Comp.
Ammoniated Tincture of Guaiacum 30 mins.; Mucilage of Acacia 1 drm.; water to 1 oz. *London Ophthalmic.*

Mistura Hæmatoxyli.
Lime Water 3 drms.; Decoction of Logwood to 1 oz. *Consumption.*

Mistura Hæmatoxyli Co.
Extract of Logwood 10 grains.; Wine of Opium 5 mins.; Ipecacuanha Wine 10 mins.; Chalk Mixture to 1 oz. *Guy's.*

Mistura Hæmatoxyli Opiata.
Tincture of Opium 3 mins.; Tincture of Catechu 1 drm.; Decoction of Logwood to 1 oz. *St. George's.*

Mistura Hydrargyri c. Cinchonâ.
Solution of Perchloride of Mercury ½ drm.; Infusion of Yellow Bark 1 oz. *Samaritan.*

Mistura Hydrargyri Iodidi.
Red Iodide of Mercury $\frac{1}{8}$ gr.; Iodide of Potassium ½ gr.; water 1 oz. *British Skin.*
Corrosive Sublimate $\frac{1}{16}$ gr.; Iodide of Potassium 4 grs.; water to 1 oz. *University.*
Solution of Perchloride of Mercury 1 drm.; Compound Mixture of Iodine to 1 oz. *Charing Cross.*
Solution of Perchloride of Mercury 30 mins.; Iodide of Potassium 3 grs.; water to 1 oz. *London Ophthalmic.*

Mistura Hydrargyri Perchloridi.
Perchloride of Mercury $\frac{1}{16}$ gr.; Tincture of Cochineal 10 mins.; water to 1 oz. *London Ophthalmic.*
Perchloride of Mercury $\frac{1}{15}$ gr.; Rectified Spirit 10 mins.; water to 1 oz. *London.*
Solution of Perchloride of Mercury 80 mins.; Infusion of Quassia to 1 oz. *Guy's.*
Solution of Perchloride of Mercury 1 drm.; Glycerine 10 mins.; water to 1 oz. *Westminster Ophthalmic.*

Mistura Hydrargyri Perchloridi Composita.
Iodide of Potassium 5 grs.; Perchloride of Mercury Mixture to 1 oz. *London.*

Mistura Hyoscyami c. Sodâ.
Tincture of Henbane 20 mins.; Bicarbonate of Soda 10 grs.; Mucilage 1 drm.; Chloric Ether 5 mins.; water to 1 oz. *Consumption.*

Mistura Iodi.
Iodine $\frac{1}{8}$ gr.; Iodide of Potassium 3 grs.; Peppermint Water to 1 oz. *Guy's.*
Tincture of Iodine 5 mins.; Iodide of Potassium 2½ grs.; Peppermint Water to 1 oz. *London.*

Mistura Iodi Co.
Tincture of Iodine 8 mins. : Iodide of Potassium 5 grs.; Cinnamon Water to 1 oz. *Charing Cross.*

Mistura Ipecacuanhæ Ammoniata.
Ipecacuanha Wine 10 mins.; Carbonate of Ammonia 5 grs.; Acacia Mixture to 1 oz. *London.*

Mistura Ipecacuanhæ c. Ammoniâ.
Ipecacuanha Wine 5 mins.; Carbonate of Ammonia 5 grs.; Camphor Water to 1 oz. *Consumption.*

Mistura Ipecacuanhæ Co.
Ipecacuanha Wine 10 mins.; Carbonate of Ammonia 4 grs.; Bicarbonate of Soda 15 grs.; Compound Tincture of Camphor 15 mins.; Chloroform Water 2 drms.; Water to 1 oz. *Charing Cross.*
Wine of Ipecacuanha 15 mins.; Tincture of Squill 7½ mins.; Tincture of Opium 4 mins.; Glycerine 18 mins.; water to 1 oz. *University.*

Mistura Ipecacuanhæ c. Hyoscyamo.
Tincture of Henbane 20 mins.; Spirit of Chloroform 10 mins.; Carbonate of Ammonia 5 grs.; Wine of Ipecacuanha 5 mins.; Camphor Water to 1 oz. *Consumption.*

Mistura Ipecacuanhæ Opiata.
Compound Tincture of Camphor 30 mins.; Wine of Ipecacuanha 10 mins.; Dill Water to 1 oz. *Consumption.*

Mistura Ipecacuanhæ c. Opio.
Ipecacuanha Wine 10 mins.; Tincture of Opium 4 mins.; water 1 oz. *British Skin.*

Mistura Ipecacuanhæ c. Sodâ.
Bicarbonate of Soda 10 grs.; Ipecacuanha Wine 20 mins.; Spirit of Nitrous Ether 15 mins.; Syrup 1 drm.; water to 1 oz. *Gt. Northern.*

Mistura Jalapæ c. Potassæ Nitrate.
Jalap in powder ½ oz.; Nitrate of Potash in powder ¼ oz.; mix. Direction: "To be mixed in a pint-and-a-half bottle of water. Shake well, and take 2 tablespoonfuls three times a day. *Royal Free.*

Mistura Krameriæ.
Ipecacuanha Wine 20 mins.; Tincture of Catechu 20 mins.; Infusion of Rhatany Root to 1 oz. *Guy's.*

Mistura Liquoris Potassæ.
Solution of Potash 20 mins.; water 1 oz. *British Skin.*

Mistura Lobeliæ.
Etherial Tincture of Lobelia 15 mins.; Camphor Water to 1 oz. *London.*

Mistura Magnesiæ.
Carbonate of Magnesia 15 grs.; water 1 oz. *Guy's.*
Carbonate of Magnesia 15 grs.; Acacia Mixture to 1 oz. *London.*

Mistura Magnesiæ Composita.
Sulphate of Magnesia 60 grs.; Carbonate of Magnesia 20 grs.; Peppermint Water 1 oz. *Charing Cross.*

Mistura Magnesiæ et Magnes. Sulph.
Carbonate of Magnesia 10 grs.; Sulphate of Magnesia 60 grs.; Peppermint Water 1 oz. *Guy's.*
Carbonate of Magnesia 15 grs.; Sulphate of Magnesia 30 grs.; Peppermint Water 1 oz. *St. Thomas's.*

Mistura Magnesiæ c. Rheo.
Compound Rhubarb Powder 20 grs.; water 1 oz. *King's.*
Rhubarb 5 grs.; Carbonate of Magnesia 15 grs.; Peppermint Water to 1 oz. *London.*
Rhubarb 5 grs.; Carbonate of Magnesia 15 grs.; Cinnamon Water 1 oz. *London Ophthalmic.*
Rhubarb 7½ grs.; Carbonate of Magnesia 15 grs.; Peppermint Water 1 oz. *St. Thomas's.*

Mistura Magnesiæ c. Sennâ.
Sulphate of Magnesia 120 grs.; Strong Tincture of Ginger 5 mins.; Infusion of Senna to 1 oz. *Charing Cross.*

Mistura Magnesiæ Sulphatis (Mistura Alba).
Sulphate of Magnesia 120 grs.; Carbonate of Magnesia 20 grs.; Peppermint Water to 2 oz. For a dose. *St. Mary's.*

Sulphate of Magnesia 60 grs.; Peppermint Water to 1 oz. *London.*

Mistura Magnesiæ Sulphatis et Acidi Gallici.
Sulphate of Magnesia 1 drm.; Gallic Acid 10 grs.; Dil. Sulphuric Acid 5 mins.; water to 1 oz. *Royal Chest.*

Mistura Magnesiæ Sulphatis Comp.
Sulphate of Magnesia 60 grs.; Carbonate of Magnesia 15 grs.; Peppermint Water to 1 oz. *London.*

Mistura Magnes. Sulph. et Magnes. Carb.
Carbonate of Magnesia 10 grs.; Sulphate of Magnesia 1 drm.; Peppermint Water to 1 oz. *Royal Chest.*

Mistura Magnesiæ Sulphatis c. Quiniâ.
Sulphate of Magnesia 60 grs.; Diluted Sulphuric Acid 4 mins.; Sulphate of Quinia 1 gr.; Syrup of Ginger 1 drm.; Dill Water to 1 oz. *Samaritan.*

Mistura Menthæ Acida.
Dil. Sulphuric Acid 11 mins.; Confection of Roses 1 drm.; Mint Water 1 oz. *St. George's.*

Mistura Menthæ Sulphurica.
Dil. Sulphuric Acid 15 mins.; Sulphate of Magnesia 1 drm.; Spearmint Water to 1 oz. *Royal Free.*

Mistura Menthæ S. c. Magn. Sulph. (S. B. H.).
Sulphate of Magnesia 60 grs.; Dil. Sulphuric Acid 10 mins.; Syrup of Red Poppy ½ drm.; Spearmint Water to 1 oz. *Samaritan.*

Mistura Morphiæ c. Acido Hydrocyan.
Hydrochlorate of Morphia $\frac{1}{6}$ gr.; Dil. Hydrocyanic Acid 4 mins.; Syrup 1 drm.; Mucilage to 1 oz. Dose ½ to 1 oz. *Consumption.*

Mistura Mucilaginosa.
Almond Oil 10 mins.; Syrup 30 mins.; Mucilage of Acacia 2½ drms.; water to 1 oz. *Guy's.*

Mistura Neuralgica.
Carbonate of Ammonia 5 grs.; Chloride of Ammonium 20 grs.; Peppermint Water 7 drms.; Mucilage 1 drm. *Samaritan.*

Mistura Nigra.
Sulphate of Magnesia 1 drm.; Essence of Peppermint 5 mins.; Comp. Infusion of Senna 1 oz. *St. Thomas's.*

Mistura Nucis Vomicæ.
Tincture of Nux Vomica 10 mins.; Comp. Infusion of Gentian 1 oz. *London Ophthalmic. Westminster Ophthalmic.*
Tincture of Nux Vomica 20 mins.; water to 1 oz. *British Skin.*

Mistura Nucis Vomicæ Acida.
Tincture of Nux Vomica 7½ mins.; Diluted Nitric Acid 10 mins.; water to 1 oz. *University.*

Mistura Nucis Vomicæ c. Ammoniâ.
Tincture of Nux Vomica 5 mins; Carbonate of Ammonia 4 grs.; Chloric Ether 5 mins.; Infusion of Quassia to 1 oz. *Consumption.*
Tincture of Nux Vomica 5 mins.; Carbonate of Ammonia 4 grs.; Chloroform Water 120 mins.; Infusion of Quassia to 1 oz. *Charing Cross.*

Mistura Olei.
Olive Oil 1 drm.; Solution of Potash 4 mins.; water to 1 oz. *Guy's.*

Mistura Olei c. Æthere.
Ether 1 drm.; Cod-Liver Oil to 6 drms. Dose 2 drms. *Consumption.*

Mistura Olei Morrhuæ.
Cod-Liver Oil 1 drm.; Lime Water 1 drm. For a dose. *St. Mary's.*
Cod-Liver Oil 30 mins.; Compound Tincture of Gentian 5 mins.; Lime Water to 1 drm. *University.*

Mistura Olei Morrhuæ c. Æthere.
Ether 80 mins.; Cod-Liver Oil to 1 oz. *Charing Cross.*

Mistura Olei Morrhuæ Amara.
Cod-Liver Oil 1 drm.; Tincture of Calumba 15 mins.; Lime Water 1 drm. For a dose. *St. Mary's.*

Mistura Olei Morrhuæ c. Ferro.
Cod-Liver Oil 1 oz.; Carbonate of Potash $\frac{1}{4}$ gr.; Wine of Iron 1 oz. Dose 2 drms. *St. Mary's.*

Mistura Olei Morrhuæ et Ferri.
Cod Liver Oil 4 drms.; Tartarated Iron 10 grs.; Solution of Potash 3 mins.; water to 1 oz. *Guy's.*

Mistura Olei Morrhuæ c. Potassâ.
Cod-Liver Oil 6 drms.; Solution of Potash 40 mins.; Stronger Solution of Ammonia 2 mins.; Oil of Cassia 1 min.; Syrup 2 drms. Dose 2 drms. *Consumption.*

Mistura Olei Morrhuæ Præparata.
Cod-Liver Oil 6 drms.; Stronger Solution of Ammonia 2 mins.; Oil of Cassia 1 min.; Syrup 2 drms. Dose 2 drms. *Consumption.*

Mistura Olei Ricini. (*See also* p. 263.)
Castor Oil $\frac{1}{2}$ oz.; Tincture of Opium 5 mins.; Powder of Acacia *q. s.*; Cinnamon Water to 1 oz. *King's.*
Castor Oil 40 mins.; Mucilage 2 drms.; Syrup 160 mins.; water to 1 oz. *Gt. Northern.*
Castor Oil 60 mins.; Tincture of Rhubarb 60 mins.; Tincture of Quillaia 20 mins.; Pimento Water to 1 oz. *Guy's.*

Mistura Oleosa.
Oil of Almonds 40 mins.; Solution of Potash 5 mins.; Oil of Caraway $\frac{1}{6}$ min.; water to 1 oz. *London.*

Mistura Olibani.
Olibanum 20 grs.; Honey 30 mins.; Barley Water to 1 oz. *Guy's.*

Mistura Opii Acida.
Tincture of Opium 5 mins.; Aromatic Sulphuric Acid 10 mins.; Syrup of Tolu 60 mins.; water to 1 oz. *Guy's.*

Mistura Oxymellis.
Oxymel 3 fl. drms.; Nitrate of Potash 7 grs.; water to 1 oz. *Guy's.*

Mistura Oxymellis Co.
Vinegar 10 mins.; Honey 24 grs.; Nitrate of Potash 10 grs.; water to 1 oz. *St. Thomas's.*
Ipecacuanha Wine 5 mins.; Comp. Tincture of Camphor 20 mins.; Oxymel Mixture to 1 oz. *Guy's.*

Mistura Oxymellis Scillæ.
Oxymel of Squill 80 mins.; Spirit of Nitrous Ether 20 mins.; Paregoric 20 mins.; water to 1 oz. *St. George's.*

Mistura Pareira Composita.
Bicarbonate of Potash 20 grs.; Tincture of Hyoscyamus 30 mins.; Decoction of Pareira to 1 oz. *Guy's.*

Mistura Pectoralis.
Ipecacuanha Wine 5 mins.; Comp. Tincture of Camphor 15 mins.; Spirit of Nitrous Ether 30 mins.; Mucilage 30 mins.; water to $\frac{1}{2}$ oz. *Samaritan.*
Ipecacuanha Wine 15 mins.; Tincture of Opium 3 mins.; Mucilage 2 drms.; water to 1 oz. *Royal Free.*

Mistura Phosphorica.
Hypophosphite of Lime 10 grs.; Syrup of Phosphate of Iron $\frac{1}{2}$ drm.; Infusion of Gentian to 1 oz. *Royal Chest.*

Mistura Plumbi Acetatis.
Acetate of Lead 2 grs.; Dil. Acetic Acid 15 mins.; water to 1 oz. *Royal Free.*
Acetate of Lead 2 grs.; Vinegar 20 mins.; water to 1 oz. *Consumption.*

Mistura Potassæ.
Solution of Potash 15 mins.; Acacia Mixture to 1 oz. *London.*
Sulphate of Potash 20 grs.; Solution of Potash 10 mins.; Peppermint Water 1 oz. *St. Thomas's.*

Mistura Potassæ Acetatis Co.
Acetate of Potash 15 grs.; Spirit of Juniper 9 mins.; Decoction of Broom to 1 oz. *University.*
Acetate of Potash 20 grs.; Nitrate of Potash 5 grs.; Spirit of Nitrous Ether $\frac{1}{2}$ drm.; Vinegar of

Squill 10 mins.; Solution of Acetate of Ammonia 3 drms.; Camphor Water to 1 oz. *Royal Chest.*

Acetate of Potash 20 grs; Tincture of Squill 15 mins.; Spirit of Nitrous Ether 30 mins.; Scoparium Juice 60 mins.; water to 1 oz. *Guy's.*

Acetate of Potash 30 grs.; Spirit of Nitrous Ether 30 mins.; Camphor Water to 1 oz. *Consumption.*

Mistura Potass. Alkal. et Mist. Potass. Alk. Eff.

Bicarbonate of Potash 22 grs.; water ½ oz. Neutralised by 15 grs. Tartaric Acid. *Samaritan.*

Mistura Potassæ Bicarbonatis.

Bicarbonate of Potash 15 grs.; Cinnamon Water to 1 oz. *Royal Free.*

Bicarbonate of Potash 20 grs.; Acetate of Potash 10 grs.; Camphor Water 1 oz. *St. Mary's.*

Bicarbonate of Potash 20 grs.; Tincture of Henbane 20 mins.; Camphor Water to 1 oz. *University.*

Bicarbonate of Potash 20 grs.; water ½ oz. *King's.*

Bicarbonate of Potash 20 grs.; Spirit of Chloroform 10 mins.; Red Mixture to 1 oz. *London.*

Bicarbonate of Potash 30 grs.; Peppermint Water 1 oz. *Charing Cross.*

Mistura Potassæ c. Calumbâ.

Bicarbonate of Potash 20 grs.; Aromatic Spirit of Ammonia 20 mins.; Spirit of Chloroform 10 mins.; Infusion of Calumba to 1 oz *London.*

Mistura Potassæ Chloratis.

Chlorate of Potash 7½ grs.; Hydrochloric Acid 2 mins.; water 1 oz. *St. Thomas's.*

Chlorate of Potash 10 grs.; water 1 oz. *Charing Cross.*

Chlorate of Potash 10 grs.; Juice of Conium 1 drm.; Peppermint Water to 1 oz. *Samaritan.*

Chlorate of Potash 10 grs.; Decoction of Cinchona 1 oz. *Guy's.*

Chlorate of Potash 10 grs.; Red Mixture 1 oz. *London.*

Chlorate of Potash 15 grs.; Dil. Hydrochloric Acid 10 mins.; water to 1 oz. *Royal Free.*

Chlorate of Potash 15 grs.; water 1 oz. *Great Northern.*

Chlorate of Potash 20 grs.; water 1 oz. *Westminster Ophthalmic.*

Mistura Potassæ Citratis. (*See also* p. 264.)
   Bicarbonate of Potash 20 grs.; Lemon Juice 3 drms.; water to 1½ ; *St. Thomas's*.
   Bicarbonate of Potash 20 grs.; Citric Acid 18 grs.; Syrup 3 drms.; water to 1½ oz. *King's*.
   Bicarbonate of Potash 22 grs.; Citric Acid 15 grs.: water to 1 oz. *University*.
   Citrate of Potash 20 grs.; water 1 oz. *London. Guy's*.
   Citrate of Potash 30 grs.; water to 1 oz. *Royal Free*.
   Citrate of Potash 30 grs.; Syrup 30 mins.; water to 1 oz. *Throat*.

Mistura Potassæ Citratis Effervescens.
   Bicarbonate of Potash 20 grs.; Lemon Juice 3 drms.; water to 1½ oz. *St. Thomas's*.
   Bicarbonate of Potash 20 grs.; water to 1 oz., with ½ oz. Citric Acid Solution in effervescence. *London*.

Mistura Potassæ Co.
   Acetate of Potash 10 grs.; Bicarbonate of Potash 10 grs.; Nitrate of Potash 5 grs.; water to 1 oz. *London*.

Mistura Potassæ Effervescens.
   Bicarbonate of Potash 20 grs.; Citric Acid 15 grs.; water to 1 oz. *Consumption*.
   Bicarbonate of Potash 22 grs.; water 1 oz.; Citric Acid 15 grs.; water 1 oz.; dissolve and mix. *University*.
   Bicarbonate of Potash 22 grs.; water to 1 oz.: effervescing with 4 drms. Lemon Juice. *Guy's*.

Mistura Potassæ Liquoris.
   Solution of Potash 20 mins.; Compound Tincture of Cardamoms 20 mins.; Caraway Water 4 drms.; water to 1 oz. *Consumption*.

Mistura Potassæ Nitratis.
   Nitrate of Potash 10 grs.; Camphor Water 1 oz. *Charing Cross*.
   Nitrate of Potash 10 grs.; Solution of Acetate of Ammonia 1 drm.; Spirit of Nitrous Ether 15 mins.; Camphor Water to 1 oz. *St. Mary's*.
   Nitrate of Potash 15 grs.; Spirit of Nitrous Ether 15 mins.; Syrup of Lemons 40 mins.; Peppermint Water to 1 oz. *Guy's*.

Mistura Potassæ Tartratis Effervescens.

Bicarbonate of Potash 15 grs.; water to ½ oz. (taken whilst effervescing with Tartaric Acid 12 grs., in ½ oz. of water). *London Ophthalmic.*

Bicarbonate of Potash 20 grs.; water 1 oz. (taken with Tartaric Acid 15 grs., in ½ oz. of water). *St. Mary's.*

Bicarbonate of Potash 20 grs.; water to 1 oz. (with ½ oz. Tartaric Acid Solution in effervescence). *London.*

Mistura Potassii Bromidi.

Bromide of Potassium 10 grs.; Bicarbonate of Potash 4 grs.; water 1 oz. *Westminster Ophthalmic.*

Bromide of Potassium 10 grs.; Aromatic Spirit of Ammonia 20 mins.; Camphor Water to 1 oz. *Throat.*

Bromide of Potassium 10 grs.; Spirit of Chloroform 10 mins.; water to 1 oz. *Royal Free.*

Bromide of Potassium 10 grs.; Camphor Water 3 drms.; water to 1 oz. *British Skin.*

Bromide of Potassium 10 grs.; Carbonate of Ammonia 3 grs.; Tincture of Calumba 30 mins.; water 1 oz. *London Ophthalmic.*

Bromide of Potassium 10 grs.; Red Mixture to 1 oz. *London.*

Bromide of Potassium 20 grs.; Compound Tincture of Cardamoms 10 mins.; Cinnamon Water to 1 oz. *Consumption.*

Bromide of Potassium 20 grs.; Peppermint Water 1 oz. *Charing Cross.*

Bromide of Potassium 20 grs.; water 1 oz. *King's.*

Mistura Potassii Bromidi Co.

Bromide of Potassium 10 grs.; Ammoniated Tincture of Valerian 10 mins.; Camphor Water to 1 oz. *Royal Chest.*

Mistura Potassii Bromidi et Iodidi.

Bromide of Potassium 10 grs.; Iodide of Potassium 5 grs.; Water to 1 oz. *London.*

Mistura Potassii Iodidi.

Iodide of Potassium 3 grs.; Comp. Infusion of Gentian 1 oz. *St. Thomas's.*

Iodide of Potassium 3 grs.; Bicarbonate of Potash 5 grs.; water to 1 oz. *Consumption. St. Mary's.*

Iodide of Potassium 3 grs.; Infusion of Quassia 1 oz. *University.*
Iodide of Potassium 3 grs.; Carbonate of Ammonia 5 grs.; Burnt Sugar 1 min.; Chloroform Water 2 drms.; Infusion of Quassia to 1 oz. *Throat.*
Iodide of Potassium 5 grs.; Red Mixture 1 oz. *London.*
Iodide of Potassium 5 grs.; water 1 oz. *Westminster Ophthalmic. British Skin.*
Iodide of Potassium 5 grs.; Infusion of Quassia 1 oz. *King's.*
Iodide of Potassium 5 grs.; Aromatic Spirit of Ammonia 15 mins.; water 1 oz. *Gt. Northern.*
Iodide of Potassium 5 grs.; Tincture of Iodine 5 mins.; water ½ oz. *Samaritan.*
Iodide of Potassium 5 grs.; Bicarbonate of Potash 10 grs.; Infusion of Quassia to 1 oz. *Royal Chest.*
Iodide of Potassium 5 grs.; Aromatic Spirit of Ammonia 20 mins.; Comp. Infusion of Gentian 4 drms.; water to 1 oz. *Guy's.*
Iodide of Potassium 5 grs.; Carbonate of Ammonia 3 grs.; Peppermint Water 1 oz. *Charing Cross.*
Iodide of Potassium 5 grs.; Carbonate of Ammonia 3 grs.; Tincture of Calumba 30 mins.; water to 1 oz. *London Ophthalmic.*

Mistura Potassii Iodidi Alkalina.
Bicarbonate of Potash 15 grs.; Carbonate of Ammonia 3 grs.; Iodide of Potassium 3 grs.; Camphor Water to 1 oz. *Consumption.*

Mistura Potassii Iodidi Ammoniata.
Iodide of Potassium 5 grs.; Carbonate of Ammonia 4 grs.; Decoction of Cinchona to 1 oz. *London.*

Mistura Potassii Iodidi Aperiens.
Iodide of Potassium 3 grs.; Sulphate of Magnesia 1 drm.; Carbonate of Magnesia ½ drm.; Spirit of Nitrous Ether ½ drm.; Peppermint Water to 1 oz. *Royal Chest.*

Mistura Potassii Iodidi Co.
Iodide of Potassium 4 grs.; Iodine $\frac{1}{20}$ gr.; water 1 oz. *Westminster Ophthalmic.*

Mistura Potassii Iodidi c. Ferro.

Iodide of Potassium 3 grs.; Bicarbonate of Potash 15 grs.; Ammonio-Citrate of Iron 15 grs.; Pimento Water to 1 oz. *Consumption.*

Iodide of Potassium 5 grs.; Ammonio-Citrate of Iron 5 grs.; water to 1 oz. *Royal Free.*

Iodide of Potassium 5 grs.; Tartarated Iron 10 grs.; Infusion of Quassia 1 oz. *Gt. Northern.*

Mistura Potassii Iodidi c. Hydrargyro.

Iodide of Potassium 5 grs.; Solution of Perchloride of Mercury Brit. Pharm. 1 drm.; Infusion of Quassia to 1 oz. *Royal Free.*

Iodide of Potassium 5 grs.; Red Iodide of Mercury $\frac{1}{16}$ gr.; Infusion of Yellow Bark 1 oz. *Samaritan.*

Mistura Potassii Iodidi c. Stramonio.

Iodide of Potassium 3 grs.; Extract of Stramonium $\frac{1}{4}$ gr.; Extract of Liquorice 2 grs.; Chloric Ether 5 mins.; water to 1 oz. *Consumption.*

Mistura Quassiæ Acida.

Diluted Nitric Acid 5 mins.; Diluted Hydrochloric Acid 8 mins.; Infusion of Quassia $\frac{1}{2}$ oz. *Samaritan.*

Mistura Quassiæ c. Ferro.

Ammonio-Citrate of Iron 12 grs.; Infusion of Quassia 1 oz. *University.*

Tinct. of Perchloride of Iron 10 mins.; Infusion of Quassia 1 oz. *Gt. Northern.*

Solution of Perchloride of Iron 15 mins.; Spirit of Chloroform 10 mins.; Infusion of Quassia to 1 oz. *London.*

Mistura Quiniæ.

Sulphate of Quinia 1 gr.; Dil. Sulphuric Acid 2 mins.; water to 1 oz. *British Skin. Royal Chest.*

Sulphate of Quinia 1 gr.; Dil. Sulphuric Acid 2 mins.; water to $\frac{1}{2}$ oz. *Samaritan.*

Sulphate of Quinia 1 gr.; Dil. Sulphuric Acid $2\frac{1}{2}$ mins.; water 1 oz. *London. Throat.*

Sulphate of Quinia 1 gr.; Dil. Sulphuric Acid 5 mins.; water 1 oz. *King's. London Ophthalmic. Royal Free.*

Sulphate of Quinia 1 gr.; Dil. Sulphuric Acid 2
mins.; Comp. Tincture of Cardamoms ½ drm.;
water 1 oz. *St. Thomas's.*
Sulphate of Quinia 1½ gr.; Dil. Sulphuric Acid 3
mins.; water 1 oz. *Westminster Ophthalmic.*
Sulphate of Quinia 2 grs.; Dil. Sulphuric Acid 2
mins.; water 1 oz. *University.*
Sulphate of Quinia 2 grs.; Dil. Sulphuric Acid 5
mins.; water to 1 oz. *Consumption.*
Sulphate of Quinia 2 grs.; Dil. Sulphuric Acid 4
mins.; Tincture of Orange Peel 10 mins.;
Simple Syrup 30 mins.; water to 1 oz. *Guy's.*
Cinchonidine 2 grs.; Dil. Sulphuric Acid 5 mins.;
water 1 oz. *Gt. Northern.*

Mistura Quiniæ Acida.
Sulphate of Quinia 1 gr.; Dil. Nitro-hydrochloric
Acid 10 mins.; Cinnamon Water 4 drms.;
water to 1 oz. *Consumption.*

Mistura Quiniæ Alkalina.
Sulphate of Quinia 1 gr.; Carbonate of Ammonia
2 grs.; Bicarbonate of Potash 20 grs.; Tragacanth 4 grs.; Chloroform Water 2 drms.; water
to 1 oz. *Charing Cross.*

Mistura Quiniæ c. Acido Hydrobromico.
Sulphate of Quinia 1 gr.; Hydrobromic Acid, medi-
cinal, 30 mins.; water to 1 oz. *London.*

Mistura Quiniæ c. Cascarillâ.
Sulphate of Quinine 1 gr.; Dil. Sulphuric Acid 2¼
mins.; water ½ oz.; Infusion of Cascarilla to
1 oz. *Consumption.*

Mistura Quiniæ Cathartica.
Sulphate of Quinia 1 gr.; Dil. Sulphuric Acid 5
mins.; Sulphate of Magnesia 60 grs.; water 1
oz. For a dose. *London Ophthalmic.*
Sulphate of Magnesia 60 grs.; Quinine Mixture to
1 oz. *London.*

Mistura Quiniæ c. Ferro.
Sulphate of Quinia 1 gr.; Dil. Sulphuric Acid 5
mins.; Sulphate of Iron 1 gr.; water to 1 oz.
London Ophthalmic.
Sulphate of Quinia 1 gr.; Sulphate of Iron 2 grs.;
Dil. Sulphuric Acid 2 mins.; water to 1 oz.
Royal Chest.

Sulphate of Quinine 1 gr.; Sulphate of Iron 2 grs.; Diluted Sulphuric Acid 3 mins.; water to 1 oz. *Royal Free.*
Sulphate of Quinia 1 gr.; Sulphate of Iron 2 grs.; Dil. Sulphuric Acid 5 mins.; Chloroform Water 3 drms.; water to 1 oz. *Charing Cross.*
Sulphate of Quinia 1 gr.; Sulphate of Iron 2 grs.; Dil. Sulphuric Acid 1 min.; water to 1 oz. *University.*
Sulphate of Quinia 1 gr.; Sulphate of Iron 1 gr.; Dil. Sulphuric Acid 2½ mins.; water to 1 oz. *Consumption.*
Sulphate of Quinia 1½ gr.; Solution of Perchloride of Iron 10 mins.; Dil. Sulphuric Acid 3 mins.; water 1 oz. *Westminster Ophthalmic.*
Sulphate of Quinia 2 grs.; Dil. Sulphuric Acid 2 mins.; Sulphate of Iron 1 gr.; water 1 oz. *St. Thomas's.*

Mistura Quiniæ et Potassæ Chloratæ.
Sulphate of Quinia 1 gr.; Chlorate of Potash 10 grs.; Dil. Hydrochloric Acid 2 mins.; water to 1 oz. *Royal Chest.*

Mistura Quiniæ Sulphatis. *See* Mist. Quiniæ.

Mistura Quiniæ c. Valerianâ.
Tincture of Valerian 30 mins.; Chloric Ether 10 mins.; Sulphate of Quinine 2 grs.; Dil. Sulphuric Acid 5 mins.; water to 1 oz. *Consumption.*

Mistura Quiniæ Vera.
Sulphate of Quinine 1 gr.; Dil. Sulphuric Acid 5 mins.; water 1 oz. *Gt. Northern.*

Mistura Rhei c. Æthere.
Tincture of Rhubarb 1 drm.; Aromatic Spirit of Ammonia 20 mins.; Spirit of Chloroform 10 mins.; water to 1 oz. *Samaritan.*

Mistura Rhei Ammoniata.
Rhubarb 4 grs.; Carbonate of Ammonia 2 grs.; Infusion of Quassia ½ oz.; Peppermint Water to 1 oz. *King's.*

Mistura Rhei Ammoniata c. Sodâ.
Rhubarb 4 grs.; Carbonate of Ammonia 3 grs.; Bicarbonate of Soda 10 grs.; Peppermint Water to 1 oz. *Royal Chest.*

Mistura Rhei Aromatica.

Tincture of Rhubarb 1 drm.; Aromatic Powder of Chalk 8 grs.; Peppermint Water to 1 oz. *St. Thomas's.*

Rhubarb Root, in powder, 5 grs.; Compound Cinnamon Powder 8 grs.; water to 1 oz. *Royal Free.*

Mistura Rhei Co.

Powdered Rhubarb 3 grs.; Carbonate of Soda 6 grs.; Ginger 3 grs.; water 1 oz. *Westminster Ophthalmic.*

Rhubarb Powder 5 grs.; Calumba 10 grs.; Bicarbonate of Soda 10 grs.; Peppermint Water 1 oz. *Guy's.*

Rhubarb 8 grs.; Carbonate of Magnesia 15 grs.; Bicarbonate of Soda 10 grs.; Ipecacuanha ¼ gr.; Peppermint Water 1 oz. *Charing Cross.*

Mistura Rhei c. Calumbâ.

Rhubarb 5 grs.; Bicarbonate of Soda 10 grs.; Aromatic Spirit of Ammonia 30 mins.; Infusion of Calumba to 1 oz. *London.*

Mistura Rhei c. Magnesiâ.

Rhubarb 7½ grs.; Carbonate of Magnesia 15 grs.; Tincture of Ginger 30 mins.; Peppermint Water to 1 oz. *University.*

Rhubarb 15 grs.; Carbonate of Magnesia 15 grs.; Solution of Ammonia 8 mins.; Caraway Water 1 oz. *St. Mary's.*

Mistura Rhei c. Opio.

Tincture of Rhubarb 2 drms.; Aromatic Spirit of Ammonia ½ drm.; Tincture of Ginger 15 mins.; Tincture of Opium 5 mins.; Peppermint Water to 1 oz. *St. Thomas's.*

Mistura Rhei c. Potassâ.

Carbonate of Ammonia 3 grs.; Comp. Rhubarb Mixture ½ oz.; Bicarbonate of Potash Mixture to 1 oz. *Charing Cross.*

Mistura Rhei c. Sodâ.

Powdered Rhubarb 5 grs.; Bicarbonate of Soda 10 grs.; Peppermint Water 1 oz. *St. Thomas's.*

Mistura Rheumatica.
Bicarbonate of Soda 10 grs.; Bicarbonate of Potash 10 grs.; Iodide of Potassium 3 grs.; Spirit of Juniper 1 drm.; Spirit of Chloroform 10 mins.; water to 1 oz. *Samaritan.*

Mistura Rosæ Laxativa.
Sulphate of Magnesia 40 grs.; Pimento Water 2 drms.; Acid Infusion of Roses to 1 oz. *Guy's.*

Mistura Rubra.
Red Solution 20 mins.; Acacia Mixture to 1 oz. *London.*
Caramel 3 grs.; water to 1 oz. *Royal Free.*

Mistura Sabinæ.
Essential Oil of Savin 1 min.; Spirit of Nitrous Ether ½ drm.; Mucilage ½ oz.; water to 1 oz. *Samaritan.*

Mistura Salina.
Solution of Acetate of Ammonia ½ oz.; water ½ oz. *Royal Free.*
Solution of Acetate of Ammonia 2 drms.; water to 1 oz. *London.*
Solution of Acetate of Ammonia, 2 drms.; Caramel 1 min.; water 1 oz. *Throat.*
Solution of Acetate of Ammonia 2½ drms.; Nitrate of Potash 10 grs.; Camphor Water to 1 oz. *St. Thomas's.*
Solution of Acetate of Ammonia 1 drm.; Citrate of Potash 15 grs.; water 7 drms. *Westminster Ophthalmic.*
Bicarbonate of Soda 10 grs.; Tartarated Soda 20 grs.; water 1 oz. *Gt. Northern.*
Nitrate of Potash 10 grs.; Treacle 10 grs.; water 1 oz. *British Skin.*

Mistura Salina Anodyna.
Saline Mixture 1 oz.; Tincture of Opium 15 mins. *St. Thomas's.*
Saline Mixture 1 oz.; Tincture of Opium 5 mins. *London.*

Mistura Salina Antimonialis.
Antimonial Wine 20 mins.; Spirit of Nitrous Ether 30 mins.; Saline Mixture to 1 oz. *London.*

Mistura Salina Aperiens.
Sulphate of Magnesia 1 drm. ; Carbonate of Magnesia 9 grs. ; Peppermint Water to 1 oz. *University.*
Sulphate of Magnesia 1 drm. ; Solution of Acetate of Ammonia ½ oz. ; Citric Acid 2 grs. ; water to 1 oz. *Royal Free.*
Sulphate of Magnesia 20 grs. ; Saline Mixture to 1 oz. *London.*
Sulphate of Magnesia 30 grs. ; Caramel 1 min. ; water to 1 oz. *Throat.*

Mistura Salina Aperiens Fortior.
Sulphate of Magnesia 3 drms. ; Carbonate of Magnesia 30 grs. ; Peppermint Water to 1 oz. *University.*

Mistura Salina Aromatica.
Aromatic Spirit of Ammonia 15 mins. ; Spirit of Nitrous Ether 30 mins. ; Solution of Acetate of Ammonia ½ oz. ; water to 1 oz. *Royal Free.*

Mistura Salina Co.
Solution of Acetate of Ammonia 2 drms. ; Nitrate of Potash 5 grs. ; Spirit of Nitrous Ether 30 mins. ; water to 1 oz. *Consumption.*
Solution of Acetate of Ammonia 2 drms. ; Oxymel of Squill 30 mins. : Ipecacuanha Wine 7½ mins.; Comp. Tincture of Camphor 15 mins. ; Spirit of Nitrous Ether 30 mins. ;. water to 1 oz. *Samaritan.*
Nitrate of Potash 5 grs. ; Compound Tincture of Camphor 10 mins. ; Spirit of Chloroform 10 mins. ; Red Mixture to 1 oz. *London.*

Mistura Salina Pot. Tart.
Tartrate of Potash 30 grs.; Comp. Tragacanth Powder 5 grs.; water to 1 oz. *London.*

Mistura Sanguinis Exsiccati.
Defibrinated Bullock's Blood Desiccated 60 grs.; Brandy 80 mins. ; Glycerine 80 mins. ; Distilled Water at 100° F. to 1 oz. *London.*

Mistura Sarsæ.
Liquid Extract of Sarsaparilla 2 drms.; water to 1 oz. *British Skin.*

Mistura Scillæ.
  Vinegar of Squill 40 mins.; Pimento Water to 1 oz. *London.*
  Vinegar of Squill ½ drm.; Dil. Acetic Acid 10 mins.; water to 1 oz. *Royal Chest.*

Mistura Scillæ Co.
  Oxymel of Squill 30 mins.; Comp. ·Tincture of Camphor 15 mins.; Spirit of Nitrous Ether 15 mins.; Camphor Water to 1 oz. *St. Mary's.*
  Syrup of Squill 30 mins.; Ipecacuanha Wine 5 mins.; Compound Tincture of Camphor 15 mins.; water 1 oz. *Westminster Ophthalmic.*
  Tincture of Squill 15 mins.; Spirit of Nitrous Ether 30 mins.; Solution of Acetate of Ammonia 1 drm.; water to 1 oz. *London Ophthalmic.*
  Tincture of Squill 20 mins.; Comp. Tincture of Camphor 30 mins.; Spirit of Nitrous Ether 30 mins.; Solution of Acetate of Ammonia 2 drms.; Camphor Water to 1 oz. *Consumption.*
  Vinegar of Squill 15 mins.; Vinegar of Ipecacuanha 5 mins.; Compound Tincture of Camphor 20 mins.; water to 1 oz. *Throat.*

Mistura Scillæ c. Conio.
  Tincture of Squill 10 mins.; Solution of Potash 15 mins.; Juice of Hemlock 30 mins.; Camphor Water to 1 oz. *Consumption.*
  Tincture of Squill 5 mins.; Solution of Potash 15 mins.; Juice of Hemlock 30 mins.; water to 1 oz. *British Skin.*

Mistura Scillæ Fort. c. Ferro.
  Tincture of Perchloride of Iron 15 mins.; Compound Tincture of Camphor 15 mins.; Compound Squill Mixture to 1 oz. *St. Mary's.*

Mistura Scoparii Co.
  Acetate of Potash 20 grs.; Nitrate of Potash 10 grs.; Spirit of Nitrous Ether 30 mins.; Spirit of Juniper 30 mins.; Decoction of Broom to 1 oz. *Consumption.*
  Acid Tartrate of Potash 30 grs.; Nitrate of Potash 5 grs.; Tincture of Hop. 60 mins.; Decoction of Broom to 1½ oz. *St. Thomas's.*
  Acid Tartrate of Potash 30 grs.; Spirit of Nitrous Ether 30 mins.; Spirit of Juniper 30 mins.; Decoction of Broom to 1 oz. *London.*

Mistura Sedativa.
Hydrate of Chloral 20 grs.; Bromide of Potassium 15 grs.; Syrup of Tolu 30 mins.; water to 1 oz. *Royal Chest.*

Mistura Senegæ.
Carbonate of Ammonia 5 grs.; Tincture of Senega 30 mins.; Infusion of Senega ½ oz.; water to 1 oz. *Guy's.*

Mistura Senegæ Acida c. Æthere.
Dil. Sulphuric Acid 5 mins.; Ether 20 mins.; Vinegar of Squill 20 mins.; Infusion of Senega to 1 oz. *Royal Chest.*

Mistura Senegæ Ammoniata.
Carbonate of Ammonia 5 grs.; Tincture of Squill 20 mins.; Infusion of Senega to 1 oz. *Royal Chest.*

Mistura Senegæ c. Ammoniâ.
Carbonate of Ammonia 4 grs.; Spirit of Ether 10 mins.; Infusion of Senega 1 oz. *Throat.*

Mistura Senegæ Co.
Carbonate of Ammonia 4 grs.; Tincture of Squill 20 mins.; Spirit of Chloroform 8 mins.; Infusion of Senega to 1 oz. *Charing Cross.*

Carbonate of Ammonia 5 grs.; Tincture of Squill 20 mins.; Infusion of Senega to 1 oz. *Consumption.*

Carbonate of Ammonia 5 grs.; Spirit of Ether 15 mins.; Tincture of Squill 20 mins.; Infusion of Senega to 1 oz. *London.*

Carbonate of Ammonia 5 grs.; Spirit of Chloroform 15 mins.; Decoction of Senega 1 oz. *St. Mary's.*

Mistura Sennæ vel Haustus Niger.
Sulphate of Magnesia 60 grs.; Solution of Ammonia 5 mins.; Extract of Liquorice 10 grs.; Infusion of Senna 1 oz. Dose 2 oz. *St. Mary's.*

Sulphate of Magnesia 1 drm.; Infusion of Senna 1 oz. *Guy's.*

Mistura Sennæ Co. (Nigra).
Sulphate of Magnesia 120 grs.; Tincture of Senna 2 drms.; Compound Tincture of Cardamoms 30 mins.; Infusion of Senna to 1 oz. *London Ophthalmic.*

Sulphate of Magnesia 60 grs.; Essence of Peppermint 5 mins.; Comp. Infusion of Senna to 1 oz. *St. Thomas's.*

Mistura Sevi.
Mutton Suet in slices 1 oz.; water 4 oz.: boil 10 minutes, and strain; then add bruised Cinnamon 1 drm., Sugar 1 oz., New Milk 16 oz.; boil again for 10 minutes, and strain. Dose 4 to 8 oz. *Guy's.*

Mistura Sodæ c. Æthere Chlorico.
Bicarbonate of Soda 10 grs.; Chloride of Sodium 3 grs.; Chloric Æther 5 mins.; Anise Water to 1 oz. *Consumption.*

Mistura Sodæ Alkalina et Mist. Sodæ Alk. Eff.
Bicarbonate of Soda 18 grs.; water ½ oz. Neutralized by 15 grs. Tartaric Acid. *Samaritan.*

Mistura Sodæ et Calumbæ.
Bicarbonate of Soda 10 grs.; Infusion of Calumba 1 oz. *Gt. Northern.*

Mistura Sodæ Effervescens.
Bicarbonate of Soda 20 grs.; Citric Acid 15 grs.; water to 1 oz. *Consumption.*

Mistura Sodæ (Tartratis) Effervescens.
Bicarbonate of Soda 20 grs., with water 1 oz. (Tartaric Acid 18 grs.; water ½ oz.). *St. Mary's.*

Mistura Sodæ Hypophosphitis.
Hypophosphite of Soda 5 grs.; water 1 oz. *Throat.*

Mistura Sodæ et Pot. Tart.
Tartarated Soda 90 grs.; Tartrate of Potash 60 grs.; Tincture of Senna 1 drm.; water to 1 oz. Dose 1 to 1½ oz. *Consumption.*

Mistura Sodæ c. Rheo.
Bicarbonate of Soda 10 grs.; Tincture of Nux Vomica 10 mins.; Chloroform 1 min.; Infusion of Rhubarb ½ oz.; Compound Infusion of Gentian to 1 oz. *Throat.*

Mistura Sodæ Salicylatis.
Salicylate of Soda 10 grs.; Peppermint Water 1 oz. *Gt. Northern. London.*
Salicylate of Soda 15 grs.; water 1 oz. *Charing Cross.*

Mistura Sodæ Sulphitis.
Sulphite of Soda 20 grs.; Infusion of Calumba 1 oz. *Royal Chest.*
Sulphite of Soda 20 grs.; Cassia Water to 1 oz. *London.*

Mistura Sodæ Sulphocarbolatis.
Sulphocarbolate of Soda 20 grs.; Camphor Water to 1 oz. *Royal Chest.*
Sulphocarbolate of Soda 20 grs.; Cassia Water to 1 oz. *London.*

Mistura Sodii Bromidi.
Bromide of Sodium 10 grs.; Compound Tincture of Lavender 4 mins.; water 1 oz. *Westminster Ophthalmic.*

Mistura Solvens.
Chloride of Ammonium 30 grs.; Liquid Extract of Liquorice 30 mins.; water to 1 oz. *Throat.*

Mistura Spiritus Vini Gallici.
Brandy 2 oz.; Water 2 oz.; Yolk of 1 Egg; Sugar ¼ oz. *London.*

Mistura Stomachica.
Rhubarb 1¼ gr.; Ginger ¾ gr.; Gentian 4½ grs.; Bicarbonate of Soda 9 grs.; cold water 1 oz.; macerate for 24 hours. *City Chest.*
Rhubarb 20 grs.; Gentian 90 grs.; Bitter Orange Peel 30 grs.; Ginger 15 grs. (all bruised); boiling water 20 oz.; Infuse 2 hours, when cold add Chloroform 20 mins. *Throat.*
Rhubarb 5 grs.; Arom. Spirit of Ammonia 30 mins.; Comp. Tincture of Gentian 30 mins.; Peppermint Water 1 oz. *Samaritan.*
Rhubarb bruised 1⅛ gr.; Bicarbonate of Soda 9 grs.; Sliced Gentian 4½ grs.; Ginger ¾ gr.; cold water 1 oz.; digest for 15 hours. *St. Thomas's.*
Rhubarb Root sliced 40 grs.; Gentian Root sliced 3 drms.; Ginger sliced ½ drm.; boiling water 20 oz.: macerate for 4 hours, strain, and add Bicarbonate of Soda ½ oz. Dose 1 to 2 oz. *Royal Free.*
Rhubarb sliced 1½ oz.; Gentian sliced 6 oz.; Ginger bruised 1 oz.; Bicarbonate of Soda 8 oz.; boiling water 4 gals.: macerate for 4 hours and strain. Dose 1 to 1½ oz. *Gt. Northern.*

Mistura Stramonii Ætheris et Opii.

Tincture of Stramonium ½ drm.; Ether ½ drm.; Tincture of Opium 5 mins.; Camphor Water to 1 oz. *Royal Chest.*

Mistura Strychniæ.

Liquor Strychniæ 4 mins.; Infusion of Gentian 1 oz. *St. Mary's.*

Liquor Strychniæ 5 mins.; Dil. Hydrochloric Acid 10 mins.; water to 1 oz. *Royal Chest.*

Liquor Strychniæ 5 mins.; Dil. Hydrochloric Acid 10 mins.; Infusion of Chiretta to 1 oz. *Charing Cross.*

Mistura Strychniæ c. Ferro.

Liquor Strychniæ 5 mins.; Dil. Nitro-hydrochloric Acid 10 mins.; Liquor Ferri Perchloridi 10 mins.; Chloroform Water ½ oz.; water to 1 oz. *Charing Cross.*

Mistura Taraxaci.

Juice of Taraxacum 1 drm.; Dil. Nitro-hydrochloric Acid 10 mins.; Tincture of Hop 30 mins.; water to 1 oz. *Consumption.*

Juice of Taraxacum 1 drm.; Dil. Nitro-hydrochloric Acid 5 mins.; water to 1 oz. *British Skin.*

Mistura Taraxaci Acida.

Juice of Dandelion 1 drm.; Dil. Nitro-hydrochloric Acid 10 mins.; Tincture of Calumba 24 mins.; Infusion of Calumba to 1 oz. *University.*

Mistura Terebinthinæ.

Oil of Turpentine 10 mins.; Mucilage 1 drm.; Cinnamon Water to 1 oz. *King's.*

Oil of Turpentine 10 mins.; Tincture of Quillaia 10 mins.; Cassia Water to 1 oz. *London.*

Oil of Turpentine 15 mins.; Mucilage 60 mins.; Peppermint Water 2 oz. *St. Mary's.*

Oil of Turpentine 20 mins.; Mucilage 3 drms.; Cinnamon Water to 1 oz. *Consumption.*

Oil of Turpentine 20 mins.; Tincture of Capsicum 20 mins.; Compound Tincture of Lavender to 2 drms. *Samaritan.*

Oil of Turpentine 20 mins.; Tincture of Quillaia 20 mins.; Tincture of Lemon Peel 30 mins.; Water to 1 oz. *Guy's.*

Mistura Terebinthinæ Chiæ.
Etherial Solution of Chian Turpentine 15 mins.; Comp. Powder of Tragacanth 10 grs.; Sublimed Sulphur 2½ grs.; Syrup 30 mins.; water to 1 oz. *London.*

Mistura Terebinthinæ et Guaiaci.
Compound Tincture of Benzoin 15 mins.; Guaiacum Mixture ½ oz.; Turpentine Mixture to 1 oz. *London.*

Mistura Terebinthinæ Purgans.
Oil of Turpentine 4 drms.; Powdered Acacia 30 grs.; Castor Oil 2 drms.; water to 1½ oz. For a dose. *St. Mary's.*

Mistura Tinct. Camph. Co.
Comp. Tincture of Camphor 15 mins.; Ipecacuanha Wine 3 mins.; Tincture of Henbane 12 mins.; water to 1 oz. *University.*

Mistura Tolutana.
Tincture of Tolu 10 mins.; Comp. Tincture of Camphor 30 mins.; Mucilage 30 mins.; Aniseed Water to 1 oz. *Consumption.*

Tincture of Tolu 10 mins.; Comp. Tincture of Camphor 20 mins.; Mucilage of Acacia 30 mins.; Aniseed Water to 1 oz. *Charing Cross.*

Mistura Tragacanthæ Co.
Comp. Powder of Tragacanth 15 grs.; Chloric Ether 10 mins.; water to 1 oz. *Consumption.*

Mistura pro Tussi.
Nitrate of Potash 4 grs.; Spirit of Nitrous Ether ⅓ drm.; Tincture of Opium 4 mins.; Vinegar of Squill 15 mins.; Treacle 15 grs.; water to 1 oz. *Royal Free.*

Solution of Hydrochlorate of Morphia 10 mins.; Ipecacuanha Wine 5 mins.; water to 1 drm. *Samaritan.*

Ipecacuanha Wine 15 mins.; Tincture of Opium 5 mins.; Nitrate of Potash 10 grs.; Mucilage 2 drms.; water to 1 oz. *St. Thomas's.*

Mistura Uvæ-Ursi.
Powdered Bearberry Leaves 10 grs.; Bicarbonate of Soda 10 grs.; Acacia Mixture to 1 oz. *London.*

Mistura Valerianæ.
Ammoniated Tincture of Valerian 30 mins.; Camphor Water to 1 oz. *London.*

Mistura Valerianæ Co.
Compound Tincture of Valerian 24 mins.; Spirit of Chloroform 18 mins.; Comp. Decoction of Aloes 114 mins.; Infusion of Valerian to 1 oz. *St. George's.*
Ammoniated Tincture of Valerian 30 mins.; Oil of Cajuput 2 mins.; Camphor Water to 1 oz. *Consumption.*

Mistura Zinci Sulphatis.
Sulphate of Zinc 1 gr.; water 1 oz. *British Skin.*

NEBULÆ.—(SPRAY SOLUTIONS.)

Aqua Acidi Carbolici Pulverisata.
(Carbolic Acid Spray.)
Carbolic Acid, No. 2 Calvert, 3 grs.; hot water 2 fl. drms. Dissolve and add water to 1 fl. oz. Mix. To be used with a Siegle's Spray Producer. *London. Royal Chest.*

Aqua Acidi Sulphurosi Pulverisata.
(Sulphurous Acid Spray.)
Sulphurous Acid 10 mins.; water to 1 fl. oz. Mix. To be used with a Siegle's Spray Producer. *London.*

Aqua Acidi Tannici Pulverisata.
(Tannic Acid Spray.)
Tannic Acid 10 grs.; water 1 fl. oz. Mix. To be used with a Siegle's Spray Producer. *London.*

Aqua Ammonii Chloridi Pulverisata.
(Chloride of Ammonium Spray.)
Chloride of Ammonium 10 grs.; Distilled water 1 fl. oz. Mix. To be used with a Siegle's Spray Producer. *London.*

Aqua Ferri Perchloridi Pulverisata.
(Perchloride of Iron Spray).
Perchloride of Iron, Crystallised, 3 grs.; water 1 fl. oz. Mix. To be used with a Siegle's Spray Producer. *London.*

Aqua Thymol Pulverisata.
(Thymol Spray.)
The same as the Thymol Lotion 1 in 1,000. To be used with a Siegle's Spray Producer. *London.*

Aqua Zinci Chloridi Pulverisata.
(Chloride of Zinc Spray).
Chloride of Zinc 2 grs.; Distilled water 1 fl. oz. Mix. To be used with a Siegle's Spray Producer. *London.*

Nebula Acidi Carbolici.
Carbolic Acid 3 grs.; Distilled Water 1 oz. Stimulant and antiseptic. *Throat.*

Nebula Acidi Lactici.
Lactic Acid U.S.P. 30 mins.; Distilled Water to 1 oz. This remedy has been found of great service in diphtheria; it appears to have the effect of dissolving the membranous exudation. *Throat.*

Nebula Acidi Sulphurosi.
Sulphurous Acid 40 to 60 mins. to be used at a time. Stimulant and antiseptic. *Throat.*
Note.—It should be inhaled very slowly.

Nebula Acidi Tannici.
Tannic Acid 5 grs.; Distilled Water 1 oz. Astringent. *Royal Chest. Throat.*

Nebula Alkalina.
Bicarbonate of Soda 15 grs.; Borax 15 grs.; Carbolic Acid 4 grs.; Glycerine 45 mins.; water to 1 oz. For use as a nasal spray. *Throat.*

Nebula Aluminii Chloridi.
Solution of Chloride of Aluminium 3 mins.; Distilled Water to 1 oz. Astringent and antiseptic. *Throat.*

Nebula Aluminis.
Alum 8 grs.; Distilled Water 1 oz. Astringent. *Royal Chest. Throat.*

Nebula Calcis.
: Lime Water, a sufficient quantity. A resolvent in cases of diphtheria. *Throat.*

Nebula Ferro-Aluminis.
: Iron Alum 3 grs.; Distilled Water 1 oz. Astringent. *Throat.*

Nebula Ferri Perchloridi.
: Perchloride of Iron 3 grs.; Distilled Water 1 oz. Astringent. *Royal Chest. Throat.*

Nebula Ferri Sulphatis.
: Sulphate of Iron 2 grs.; Distilled Water 1 oz. Astringent. *Royal Chest. Throat.*

Nebula Iodi, cum Acido Tannico.
: Tincture of Iodine 3 mins.; Glycerine of Tannic Acid 12 mins.; Distilled Water to 1 oz. For a nasal spray, to be used night and morning. *Throat.*

Nebula Iodoformi.
: Iodoform 40 grs.; Ether, S.G. ·735 1 oz. Strongly antiseptic and detergent. *Throat.*

Nebula Potassæ Chloratis.
: Chlorate of Potash 20 grs.; Water 1 oz. Detergent. *Throat.*

Nebula Potassæ Permanganatis.
: Permanganate of Potash 5 grs.; Distilled Water 1 oz. Antiseptic. *Throat.*

Nebula Potassii Bromidi.
: Bromide of Potassium 20 grs.; Water 1 oz. Sedative. *Throat.*

Nebula Sodæ Benzoatis.
: Benzoate of Soda 20 grs.; Water 1 oz. Antiseptic. *Throat.*

Nebula Sodæ Salicylatis.
: · Salicylate of Soda 20 grs.; Water 1 oz. Useful in diphtheria. *Throat.*

Nebula Sodii Chloridi.
> Chloride of Sodium, 5 grs.; Distilled Water 1 oz. Stimulant. *Royal Chest. Throat.*

Nebula Zinci Chloridi.
> Chloride of Zinc 2 grs.; Distilled Water, 1 oz. Astringent. *Royal Chest. Throat.*

Nebula Zinci Iodati
> Iodated Zinc Caustic 2 mins., or more if prescribed; Distilled Water to 1 oz. To be used with a spray producer. *Throat.*

Nebula Zinci Sulphatis.
> Sulphate of Zinc, 5 grs.; Distilled Water, 1 oz. Astringent. *Royal Chest. Throat.*

Nebula Zinci Sulphocarbolatis.
> Sulphocarbolate of Zinc 5 grs.; Distilled Water 1 oz. *Throat.*

NITRE PAPER. *See* CHARTA, *also* FUMI.

OLEUM CARBOLATUM.

Oleum Acidi Carbolici.
> Carbolic Acid (No. 2 Calvert) 1 fl. oz.; Olive Oil to 10, 20, or 40 fl. oz. *London.*
> Carbolic Acid 24 grs.; Olive Oil 1 oz. *Guy's.*

Oleum Carbolatum.
> Carbolic Acid liquefied by heat 12 mins.; Olive Oil to 1 oz. *St. Bartholomew's.*
> Carbolic Acid 12 grs.; Olive Oil 1 oz. *Middlesex.*
> Carbolic Acid 12 grs.; Olive Oil 1 drm. *Westminster.*

PASTA.

Pasta Acidi Arseniosi.
> Arsenious Acid 20 grs.; Vermilion 100 grs.; Burnt Sponge, 45 grs. Mix thoroughly. To be made into a thick paste with Glycerine when required for use. *London.*

Pasta Amyli Iodidi.
Starch Powder 1 oz.; Glycerine 2 fl. oz.; Water 6 oz. Boil together, and when nearly cold add Liquoris Iodi P.B. 1 fl. oz. *Middlesex.*

Pasta Caustica.
Extract of Sanguinaria ½ oz.; Extract of Stramonium 1 oz.; mix, and add when wanted Chloride of Zinc ½ oz. *Guy's.*

Pasta Caustica c. Opio.
Fused Chloride of Zinc 480 grs.; Flour 180 grs.; Battley's Solution of Opium 1 fl. oz. *Guy's.*

Pasta Glycerini.
Starch 1 oz.; Glycerine 6 oz.; Water 6 oz. Mix. *Gt. Northern.*

Pasta Iodi et Amyli.
Starch 1 oz.; Glycerine 2 oz.; water 6 oz.; boil together, and when nearly cold add Solution of Iodine 1 oz. *London Ophthalmic. University.*

Pasta Iodi c. Creasoto.
Iodine 120 grs.; Creasote to 1 oz. *St. Bartholomew's.*

Pasta Londinensis.
Caustic Soda, Unslaked Lime, equal parts reduced to a fine powder, and kept in a well-stopped bottle. To be made into a paste with water when required. *London. Throat.*

Nebula Pro Tactu.
Brown Soap 2 oz.; Glycerine 2 oz.; Carbolic Acid 1 drm.; Oil of Cloves 12 drops; mix, and add (if necessary) Rectified Spirit 3 drms. *Women.*

Nebula Sodæ Benzoatis c. Chloridi.
Throat

Nebula Sodæ Chloridi.
Chloride of Zinc, 1, 2, or 4 parts.; Starch in Powder 4 parts.; Prepared Lard 1 part.; Glycerine of Starch a sufficiency. *University.*

Nebula Salicylate Zinci Chloridi c. Opio 1 oz. (*formula see next diphth*); Wheat Flour 2 drms. *Middlesex.*

(Liquor Zinci Chloridi c. Opio.
Chloride of Zinc 16 oz.; Powdered Opium 1½ oz.; Hydrochloric Acid 6 drms.; Boiling Water to 20 oz. *Middlesex.*)
Chloride of Zinc 1 oz.; Wheat Flour, 2 drms.; Sedative Solution of Opium (Liq. Opii. Sed.) 1 oz. Mix. *Gt. Northern.*
Chloride of Zinc and Flour equal parts; Glycerine *q. s.*; rub the first into a thin paste with water, then add the flour; mix well and make into a thick paste with Glycerine. *London.*

Pasta Zinci Chloridi c. Opio.
Chloride of Zinc paste 1 oz.; Extract of Opium 20 grs. Rub the Extract smooth with a few drops of water, and then mix thoroughly with the paste. *London.*

Pasta Zinci Chloridi Co.
Chloride of Zinc Paste. (1, 2, or 4 pts.) 1 oz.; Extract of Opium 20 grs.; water 10 mins. *University.*

PASTILLI.

Glyco-Gelatine.
Refined Gelatine 1 oz.; Glycerine (by weight) 2½ oz.; Ammoniacal solution of Carmine, a sufficient quantity; Orange-flower Water 2½ oz. The process to be pursued in making the basis is as follows:—Soak the Gelatine in the water for two hours, then heat in a water-bath till dissolved; add the Glycerine, and stir well together. Let the mixture cool, and when nearly cold add the Carmine solution, mix till uniformly coloured, and set aside to solidify. After medicating, as directed in the following formulæ, it is cooled by pouring into an oiled tray, and, when solidified, cut into the required number of pastils. In these formulæ directions are given for making one pastil. 1 oz. of the mass will make twenty-four. *Throat.*

Pastillus Acidi Boracici.
Boracic Acid in fine powder 2 grs.; Glycerine 2 mins.; rub together and add the mixture to the Glyco-gelatine (melted in a water-bath) 18 grs. Mix and set aside to cool, and make one pastil. In aphthous affections of the mouth and throat. *Throat.*

Pastillus Acidi Carbolici.

Carbolic Acid ½ gr.; Glyco-gelatine 18 grs.; melt the Glyco-gelatine in a water-bath, add the Carbolic Acid and dissolve; then set aside to cool, and make one pastil. Antiseptic and stimulant. *Throat.*

Pastillus Ammonii Chloridi.

Chloride of Ammonium 2 grs.; rub together and add the mixture to the Glyco-gelatine (melted in a water-bath) 18 grs.; mix and set aside to cool, and make one pastil. Let one dissolve slowly at the back of the tongue every half-hour, every hour, or every two hours. In chronic pharyngeal catarrh. *Throat.*

Pastillus Bismuthi.

Carbonate of Bismuth 3 grs.; Glycerine 3 mins.; rub together, and add the mixture to Glyco-gelatine (melted in a water bath) 18 grs. Mix and set aside to cool, and make one pastil. One every two or three hours. In congestion of the pharynx with insufficient secretion. *Throat.*

Pastillus Bismuthi et Morphiæ.

Carbonate of Bismuth 3 grs.; Acetate of Morphia $\frac{1}{16}$ gr.; Glycerine 3 mins.; rub together and add the mixture to the Glyco-gelatine (melted in a water-bath) 18 grs. Mix and set aside to cool, and make one pastil. Let one pastil slowly dissolve at the back of the tongue at intervals of one or two hours, to relieve dryness of the throat and tickling cough. In sub-acute catarrh of the pharynx and larynx; and is a useful sedative in laryngeal phthisis. *Throat.*

Pastillus Bismuthi et Potassæ Chloratis.

Carbonate of Bismuth 3 grs.; Chlorate of Potash 2 grs.; Glycerine 3 mins.; rub together, and add the mixture to the Glyco-gelatine (melted in a water-bath) 18 grs. Mix and set aside to cool, and make one pastil. Let one dissolve slowly at the back of the tongue every hour or two. In granular pharyngitis with follicular ulcerations of the tonsils, and in aphthæ of the mouth. *Throat.*

Pastillus Iodoformi.
Iodoform in fine powder 1 Gr. (more or less if prescribed); Glycerine 1 min.; rub together, and add the mixture to the Glyco-gelatine (melted in a water-bath) 18 grs. Mix and set aside to cool, and make one pastil. Let one dissolve slowly on the tongue every two, three, or four hours. In syphilitic eruptions of the tongue, mouth, and throat, and in chronic pharyngitis. *Throat.*

PEPTONISED FOODS.

Peptonised Milk.
Take a pint of milk, dilute it with a quarter of a pint of water, and divide the mixture into two equal portions; heat one portion to the boiling point, and then mix it with the cold portion in a jug provided with a cover; add to this 3 fl. drms. of Pancreatic Solution, and 20 grs. of Bicarbonate of Soda, and mix well together; set the covered jug aside in a warm situation for two hours and a half, and then boil the contents for three minutes. *London. Middlesex.*

Peptonised Gruel.
To a pint of well-boiled Gruel, made thick and strong, and allowed to cool to 140° F.,* add 4 fl. drms. of Pancreatic Solution, and mix well together; pour the mixture into a jug provided with a cover, and set aside in a warm situation for two hours; then boil the product for three minutes, and finally strain.

Peptonised Milk-Gruel.
Gruel, while still boiling hot, is added to an equal quantity of cold Milk, in a jug provided with a cover. To each pint of this mixture 3 fl. drms. of Pancreatic Solution, together with 20 grs. of Bicarbonate of Soda, are added, and the whole well stirred. The covered jug is then set

---

* *This temperature can be estimated with sufficient accuracy, should no suitable thermometer be at hand, by tasting. If too hot to sip without burning the mouth, it would entirely destroy the activity of the Liquor Pancreaticus, and must be allowed to cool before such addition is made. London. Middlesex.*

aside in a warm situation for two hours, the contents boiled for three minutes, and then strained. *London. Middlesex.*

Peptonised Beef Tea.
Mix half a pound of finely minced lean Beef with a pint of water, and 20 grs. of Bicarbonate of Soda, and simmer gently for one hour and a half. When it has cooled down to 140° F.* add 4 fl. drms. of Pancreatic Solution, and keep the mixture in a warm situation for two hours, agitating it occasionally. Then strain without pressure, and boil the strained liquor for five minutes. *London. Middlesex.*

PESSI.
(*See also* SUPPOSITORIA.)

Basis pro Pesariis.
Oil of Theobroma 1 oz.; Olive Oil 2 drms.; mix together with a gentle heat. *London.*

Gelatine Pessary Mass.
Take of pure gelatine in thin strips 1 oz.; distilled water 1 oz.; Glycerine 3½ oz. Soften the gelatine by soaking it in the water until all of the water has (within a few minutes) become absorbed by the gelatine. Dissolve the softened gelatine in the glycerine previously heated on a water-bath, and allow the solution to solidify. Keep the stiff jelly in a covered pot for use. *British Skin.*

Massa pro Pessis.
Gelatine 1, softened by steeping in water; Glycerine 4; melt together in a water-bath. *Women.*

Pessus Acidi Carbolici.
Carbolic Acid (No. 2 Calvert) 1 or 2 grs.; Basis 70 grs. *London.*

Pessus Acidi Tannici.
Tannic Acid 10 grs.; Basis 60 grs. *London.*
Tannic Acid 10 or 20 grs.; Oil of Theobroma *q. s.*; mix. *Women.*
Tannic Acid 30 grs.; Oil of Theobroma 60 grs. *British Skin.*

---

* See *ante.*

Pessus Aluminis.
Alum 15 grs.; Oil of Theobroma 2 drms. *Westminster.*
Dried Alum 6 grs.; Oil of Theobroma 20 grs. *Women.*

Pessus Aluminis et Zinci.
Dried Alum 5 grs.; Sulphate of Zinc 5 grs.; Opium in powder 1 gr.; Basis 60 grs. *London.*

Pessus Atropiæ.
Atropia $\frac{1}{20}$ gr.; Oil of Theobroma 60 grs. *Charing Cross.*

Pessus Atropiæ et Morphiæ.
Sulphate of Atropia $\frac{1}{40}$ gr.; Acetate of Morphia ½ gr.; Mass 25 grs.; mix. *Women.*

Pessus Belladonnæ.
Extract of Belladonna 2 grs.; Mass 20 grs.; mix. *Women.*
Extract of Belladonna 3 grs.; Basis 70 grs. Rub the Extract into a thin paste with a few drops of water, and mix it with the Basis previously melted. *London.*

Pessus Coniæ.
Conia ½ min.; Mass 20 grs.; mix. *Women.*

Pessus Hydrargyri.
Mercurial Ointment 30 grs.; Oil of Theobroma to 2 drms. *Westminster.*
Mercurial Ointment 3 grs.; Mass 20 grs.; mix. *Women.*
Mercurial Ointment 10 grs.; Basis 60 grs. *London.*

Pessus Hydrargyri et Belladonnæ.
Ointment of Mercury 5 grs.; Extract of Belladonna 3 grs.; Basis 60 grs. *London.*

Pessus Morphiæ.
Hydrochlorate of Morphia ½ gr.; Basis 70 grs. *London.*
Hydrochlorate of Morphia 1 gr.; Oil of Theobroma 60 grs. *Charing Cross.*

Acetate of Morphia ⅓ gr.; Gelatine Pessary Mass 90 grs. *British Skin.*
Acetate of Morphia ⅓ or ½ gr.; Mass 20 grs. *Women.*

Pessus Plumbi Iodidi et Atropiæ.
Iodide of Lead 10 grs.; Sulphate of Atropia 1/18 gr.; Basis 60 grs. *London.*

Pessus Plumbi c. Iodo.
Acetate of Lead 6 grs.; Iodine 1 gr.; Mass 20 grs.; mix. *Women.*

Pessus Potassii Bromidi et Belladonnæ.
Bromide of Potassium 10 grs.; Extract of Belladonna 1 gr.; Theobroma Oil *q. s. Women.*

Pessus Zinci.
Oxide of Zinc 15 grs.; Mass 20 grs.; mix. *Women.*

Pessus Zinci et Atropiæ.
Dried Sulphate of Zinc 10 grs.; Sulphate of Atropia 1/18 gr. Basis 60 grs. *London.*

PHENOL IODATUM.
Iodine 40 grs.; Liquefied Carbolic Acid 1 oz. *Women.*

PIGMENTA.

Pigmentum Acidi Carbolici.
Carbolic Acid 1 drm.; Olive Oil to 10 drms. *Royal Free.*
Carbolic Acid ½ oz.; Glycerine ½ oz. *British Skin.*
Carbolic Acid 30 grs.; water to 1 oz. *Throat.*

Pigmentum Acidi Carbolici Mitius.
Carbolic Acid 1 drm.; Glycerine to 1 oz. *King's.*

Pigmentum Acidi Tannici.
Tannic Acid 180 grs.; Glycerine 6 drms. Dissolve by aid of heat. *British Skin.*
Tannic Acid ½ oz.; Rectified Spirit ½ oz.; water ½ oz. *London.*

Pigmentum Aloes.
Extract of Barbadoes Aloes 180 grs.; Glycerine 6 drms. *British Skin.*

Pigmentum Aluminii Acetatis.
Potash Alum 34 grs.; Distilled Water 2 drms.; dissolve. Acetate of Lead 40 grs.; Distilled Water 2 drms.; dissolve. Mix the two solutions and filter, adding more Distilled Water over the filter till 1 oz. of filtrate is obtained. *Throat.*
This pigment may be diluted with an equal quantity of water for use as a gargle, nasal douche, or spray inhalation.

Pigmentum Aluminii Chloridi.
15 mins. of a Solution Sp. G. 1250; water to 1 oz. *Throat.*

Pigmentum Argenti Nitratis Æthereum.
Nitrate of Silver 20 grs.; Distilled Water 1 drm.; Spirit of Nitrous Ether to 1 oz. *London.*

Pigmentum Argenti Nitratis Dilutum.
Nitrate of Silver 30 grs.; water to 1 oz. *Throat.*

Pigmentum Argenti Nitratis Forte.
Nitrate of Silver 60 grs.; water to 1 oz. *Throat.*

Pigmentum Belladonnæ.
Extract of Belladonna 2 drms.; Glycerine 2 drms.; water 1 drm. *Westminster. Westminster Ophthalmic.*

Pigmentum Benzolis.
(Benzine.) *British Skin.*

Pigmentum Calcis Sulphuratæ.
Synonym.—Vlemingkx's Solution.
Slaked Lime 60 grs.; Sublimed Sulphur 90 grs.; Distilled Water 1 fl. oz. and 5 fl. drms. Heat the Sulphur and Lime (previously well mixed) in the water, stirring diligently with a slip of wood; boil until the mixture measures only 1 fl. oz., then filter. *Caution.*—This mixture needs to be used with some discretion, and on occasion should be diluted. *British Skin.*

Pigmentum Cassuvii.
Cashew-nut Oil. The oil expressed from the pericarp of the fruit of Anacardium Occidentale, or from the corresponding part of the Semecarpus Anacardium. *British Skin.*

Pigmentum Chloral et Camphoræ.
Hydrate of Chloral, Camphor equal parts. Rub together in a warm mortar until completely liquefied, and filter. *Throat.*
As an anæsthetic applied externally in neuralgic affections of the throat.

Pigmentum Chloralo-Camphoratum.
Hydrate of Chloral, Camphor equal parts. Rub together till liquefied. *London.*

Pigmentum Cupri Sulphatis.
Crystals of Sulphate of Copper 15 grs.; water to 1 oz. *Throat.*

Pigmentum Ferri Perchloridi.
Solution of Perchloride of Iron ½ oz.; Glycerine to 1 oz. *King's.*
Perchloride of Iron crystallized ¼ oz.; Glycerine ½ oz.; water to 1 oz. *London.*

Pigmentum Ferri Perchloridi Dilutum.
Crystallized Perchloride of Iron 60 grs.; water to 1 oz. *Throat.*

Pigmentum Ferri Perchloridi Forte.
Crystallized Perchloride of Iron 120 grs.; water to 1 oz. *Throat.*
Crystallized Perchloride of Iron ½ oz.; Glycerine ½ oz.; water to 1 oz. *London.*

Pigmentum Ferri Sulphatis.
Sulphate of Iron 60 grs.; water to 1 oz. *Throat.*

Pigmentum Ferro-Aluminis.
Iron Alum 60 grs.; water to 1 oz. *Throat.*

Pigmentum Glycerini.
Glycerine ½ oz.; Rectified Spirit ½ oz. *British Skin.*

Pigmentum Guttæ Perchæ.
Gutta Percha 90 grs.; Chloroform 1 oz. *Fever.*

Pigmentum Hydrargyri.
Blue Pill 1 drm.; Glycerine sufficient. *Westminster.*

Pigmentum Hydrargyri Perchloridi.
Perchloride of Mercury 2 grs. ; Glycerine 2 drms. ; Rectified Spirit 1 oz. *British Skin.*

Pigmentum Iodi.
Iodine 30 grs.; Iodide of Potassium 30 grs.; Rectified Spirit 1 oz. *Middlesex.*
Iodine 40 grs. ; Iodide of Potassium 1 drm.; Rectified Spirit 1 oz. Mix. *Gt. Northern.*
Iodine 40 grs. ; Iodide of Potassium 20 grs.; Rectified Spirit to 1 oz. *London.*
Iodine 1 drm.; Spirit of Wine 1 oz. *King's.*
Iodine 80 grs. ; Iodide of Potassium 40 grs.; Rectified Spirit 1 oz. *Fever.*
Iodine 2 drms.; Iodide of Potassium 2 drms.; Methylated Spirit 4 drms. ; water to 1 oz. *Royal Free.*
Iodine $\frac{1}{2}$ oz.; Iodide of Potassium 120 grs. ; Glycerine 1 oz. *British Skin.*
Liniment of Iodine 2 drms. ; Rectified Spirit 6 drms. *Westminster. Westminster Ophthalmic.*
Linimentum Iodi P. B. *Throat.*

Pigmentum Iodi et Olei Picis.
Iodine 2 drms. ; Light Oil of Wood Tar 1 oz. : mix carefully, apply heat if necessary ; after ebullition preserve for use. *University.*

Pigmentum Olei Cadi.
Juniper Tar (Huile de Cade) *q. s.* *British Skin.*

Pigmentum Olei Ricini.
Castor Oil 3 drms. ; Rectified Spirit 6 drms. *British Skin.*

Pigmentum Olei Ricini c. Collodio.
Collodion 4 drms.; Castor Oil to $1\frac{1}{2}$ oz. *Fever.*
Collodion 4 drms. ;. Castor Oil to 1 oz. *King's.*

Pigmentum Ovi.
The Whites of 2 eggs ; Rectified Spirit 1 oz. *Fever.*

Pigmentum Picis.
Tar 1 fl. oz.; Methylated Spirit 1 oz. *Royal Free.*

Pigmentum Picis Fortius.
Tar $\frac{1}{2}$ oz.; Rectified Spirit to 1 oz. *King's.*

Pigmentum Picis c. Iodo. (*Coster's Paste.*)
Iodine 120 grs.; Rectified Oil of Tar* 1 oz. Dissolve cautiously. *British Skin.*

Pigmentum Picis Ligni Fossilis. (Coal-tar Solution).
Soap Bark bruised 4 oz.; Coal-tar 5 oz.; Proof Spirit 20 oz.
Percolate the Soap Bark with the Proof Spirit, press, filter, and make up to 20 fl. oz.
Heat the Tincture of Soap Bark and the Coal-tar together, for an hour, in a covered vessel placed in a water-bath, agitating occasionally. Then remove the vessel and stir till cold—finally strain.

Pigmentum Picis Liquidæ.
Tar ½ oz.; Rectified Spirit ½ oz. *British Skin.*

Pigmentum Picis Mitius.
Tar 1 drm.; Rectified Spirit 7 drms. *King's.*

Pigmentum Plumbi. (Glycerole of Subacetate of Lead.)
Acetate of Lead 120 grs.; Litharge in powder 84 grs.; Glycerine 1 fl. oz.; Digest the Acetate of Lead and the Litharge in the Glycerine (heated to 300° in an Oil-bath) for half an hour, constantly stirring. Then filter in a chamber heated to 300°. (Diluted to a varying extent with Glycerine, most commonly with seven times or with three times its quantity of Glycerine) as an astringent and sedative in cases of Chronic Eczema. *British Skin.*

Pigmentum Sulphuris.
Precipitated Sulphur, Carbonate of Potash, Glycerine, Rectified Spirit, equal parts. *King's.*

Pigmentum Tolutanum.
Balsam of Tolu 96 grs.; Ether (*S. G.* ·735) 1 oz.; Dissolve. *Throat.*

Pigmentum Zinci Chloridi Dilutum.
Chloride of Zinc 15 grs.; water to 1 oz. *Throat.*

Pigmentum Zinci Chloridi Forte.
Chloride of Zinc 30 grs.; water to 1 oz. *Throat.*

---

* A colourless volatile liquid distilled from Tar *s.g.* ·86.

Pigmentum Zinci Sulphatis.
Sulphate of Zinc 60 grs. ; water to 1 oz. *Throat.*

PILULÆ.

Pilula Acidi Arseniosi. (Asiatic Pill.)
Arsenious Acid $\frac{1}{12}$ gr. ; Black Pepper $\frac{1}{2}$ gr. ; Extract of Gentian 2 grs. *British Skin.*

Pilula Acidi Carbolici.
Carbolic Acid (No. 1 Calvert) $\frac{1}{4}$, $\frac{1}{3}$, $\frac{1}{2}$, or 1 gr.; Liquorice Root powdered 2 grs. ; Confection of Hips, *q. s.* in one pill, varnish. *London.*

Pilula Acidi Carbolici Composita.
Carbolic Acid 1 gr. ; Guaiacum Resin 1 gr.; Powdered Rhubarb 2 grs. *Guy's.*

Pilula Acidi Gallici.
Gallic Acid 2½ grs.; Extract of Henbane 1 gr. *Royal Chest.*
Gallic Acid 3 grs. ; Extract of Rhatany 2 grs. ; Glycerine *q. s.*: in one pill. *Samaritan.*
Gallic Acid 3½ grs. ; Extract of Gentian 1½ gr. ; in one pill. Dose 1 or 2. *Consumption.*
Gallic Acid 4 grs. ; Confection of Roses *q. s.*: in one pill. *City Chest.*
Gallic Acid 5 grs.; Glycerine 1 min.: in one pill. *London.*

Pilula Acidi Gallici c. Morphiâ.
Gallic Acid 3½ grs.; Hydrochlorate of Morphia $\frac{1}{8}$ gr. ; Extract of Gentian *q. s*.: in one pill. Dose 1 or 2. *Consumption.*

Pilula Acidi Gallici et Opii.
Gallic Acid 4 grs.; Extract of Opium $\frac{1}{4}$ gr.; Confection of Roses *q. s.*: in one pill. *City Chest.*
Gallic Acid Pill 5 grs. ; Opium in Powder $\frac{1}{2}$ gr. : in one pill. *London.*

Pilula Acidi Tannici c. Opio.
Tannic Acid 4 grs.; Opium $\frac{1}{4}$ gr.; Mucilage of Tragacanth *q. s.*: in one pill. *City Chest.*
Tannic Acid 5 grs.; Powder of Opium $\frac{1}{4}$ gr.; Mucilage to form one pill. *St. Thomas's.*

Pilula Aconiti et Belladonnæ.
Extract of Aconite ⅙ gr.; Extract of Belladonna ⅙ gr.; Sugar of Milk 2 grs.; Treacle *q. s. London Ophthalmic.*

Pilula Aloes et Assafœtidæ c. Ferro.
Sulphate of Iron 1 gr.; Pill of Aloes and Assafœtida 4 grs. *British Skin. Women.*

Pilula Aloes c. Belladonnâ.
Extract of Aloes 1 gr.; Extract of Belladonna ⅓ gr. *St. Mary's.*
Extract of Socotrine Aloes 1 gr.; Extract of Belladonna ¼ gr. *Charing Cross.*
ExBtract of arbadoes Aloes 1 gr.; Resin of Jalap ½ gr.; Extract of Belladonna ⅛ gr.; Confection of Roses to 3 grs. *Middlesex.*
Socotrine Aloes 1½ gr.; Extract of Belladonna ¼ gr. *Consumption.*
Extract of Socotrine Aloes 1½ gr.; Extract of Nux Vomica ½ gr.; Extract of Belladonna ¼ gr. *Throat.*

Pilula Aloes Composita.
Socotrine Aloes 1 gr.; Ginger 1 gr.; Extract of Gentian 2 grs.: in one pill. *London.*

Pilula Aloes c. Ferro.
Extract of Socotrine Aloes ½ gr.; Dried Sulphate of Iron 1½ gr.; Extract of Gentian 2 grs.: in one pill. *London.*
Aloes 1 gr.; Dried Sulphate of Iron 1 gr.; Extract of Gentian 2 grs.: in one pill. *Throat.*
Socotrine Aloes 1 gr.; Sulphate of Iron 1 gr.; Extract of Gentian 2 grs.; Capsicum ⅓ gr.; Treacle *q. s.*: for one pill. *Westminster Ophthalmic.*
Socotrine Aloes 2 grs.; Sulphate of Iron 1 gr.; Hard Soap 2 grs. *Gt. Northern.*
Pill of Aloes and Assafœtida 4 grs.; Sulphate of Iron 1 gr. *Fever.*

Pilula Aloes et Ferri c. Quiniâ.
Sulphate of Quinine 1 gr.; Aloes with Iron pill 4 grs.: in one pill. *London.*

Pilula Aloes c. Ipecacuanhâ.

Extract of Socotrine Aloes 1 gr.; Ipecacuanha $\frac{1}{2}$ gr.; Extract of Gentian $1\frac{1}{2}$ gr.: in one pill. Dose 1 or 2. *Consumption.*

Pilula Aloes c. Nuce Vomicâ.

Extract of Socotrine Aloes 1 gr.; Extract of Nux Vomica $\frac{1}{4}$ gr. *Charing Cross.*

Extract of Aloes 1 gr.; Extract of Nux Vomica $\frac{1}{4}$ gr.; Myrrh 2 grs.; Soap *q. s.*: in one pill. *City Chest.*

Extract of Aloes 1 gr.; Extract of Nux Vomica $\frac{1}{8}$ gr.; Soap *q. s.* *St. Mary's.*

Extract of Socotrine Aloes 1 gr.; Extract of Henbane 2 grs.; Extract of Nux Vomica $\frac{1}{4}$ gr. *Middlesex. Women.*

Extract of Socotrine Aloes 1 gr.; Extract of Henbane 1 gr.; Extract of Nux Vomica $\frac{1}{2}$ gr.; Soap $1\frac{1}{2}$ gr.: to make one pill. *St. Thomas's.*

Extract of Aloes $1\frac{1}{2}$ gr.; Extract of Nux Vomica $\frac{1}{2}$ gr.; Extract of Belladonna $\frac{1}{8}$ gr.; Powdered Liquorice 2 grs.; Glycerine of Tragacanth *q. s.*: for one pill. *University.*

Extract of Socotrine Aloes 2 grs.; Extract of Nux Vomica $\frac{1}{4}$ gr. *Consumption.*

Extract of Barbadoes Aloes 2 grs.; Extract of Hyoscyamus 1 gr.; Extract of Nux Vomica $\frac{1}{4}$ gr. *Westminster.*

Pill of Barbadoes Aloes 4 grs.; Extract of Nux Vomica $\frac{1}{4}$ gr. *Royal Chest.*

Pill of Barbadoes Aloes $4\frac{1}{2}$ grs.; Extract of Nux Vomica $\frac{1}{4}$ gr.; Treacle *q. s.*: for one pill. *St. Bartholomew's.*

Pilula Aloes c. Sapone.

Cape Aloes 2 grs.; Hard Soap 2 grs.; water *q. s.*: in one pill. *St. Bartholomew's.*

Pilula Alterativa.

Compound Extract of Colocynth 1 gr.; Extract of Henbane 1 gr.; Blue Pill 1 gr.; Ipecacuanha $\frac{1}{4}$ gr.: in one pill. *Samaritan.*

Pilula Antimonii Opiata.

Tartar Emetic $\frac{1}{4}$ gr.; Opium $\frac{1}{2}$ gr.; Treacle *q. s.*: in one pill. *Guy's.*

Pilula Aperiens.

Barbadoes Aloes, Scammony, Jalap, Ginger, of each 1 gr.; Treacle sufficient to make one pill. *London Ophthalmic. Westminster Ophthalmic.*

Gamboge 120 grs.; Soft Soap 1 oz.; Barbadoes Aloes 4 oz.; Jalap 4 oz.; Colocynth 1½ oz.; Ginger 60 grs.; Olive Oil 2 drms.; Oil of Cloves 1 drm.; Treacle *q. s.* : mix. Dose 5 to 10 grs. *Gt. Northern.*

Pilula Aperiens Fortior.

Socotrine Aloes 2 grs.; Colocynth 1 gr.; Gamboge 1 gr.; Oil of Cassia ½ min.; Compound Decoction of Aloes *q. s.* : in one pill. *City Chest.*

Pilula Argenti Nitratis.

Nitrate of Silver ⅙ gr.; Bread Mass *q. s. Royal Chest.*

Nitrate of Silver ⅙ gr.; Dover's Powder 2 grs.; Mucilage *q. s.* : in one pill. *City Chest.*

Nitrate of Silver ¼ gr.; water *q. s.*; Flour 2 grs. : in one pill. Dose 1 or 2. *Consumption.*

Nitrate of Silver ½ gr.; Extract of Gentian 1⅔ gr.; Liquorice Root in powder 1⅔ gr.: in one pill. *London.*

Nitrate of Silver ½ gr.; Opium ¼ gr.; Extract of Henbane 1 gr. *Fever.*

Pilula Argenti Oxidi.

Oxide of Silver ½ gr.; Extract of Hops 2 grs. : in one pill. *City Chest.*

Oxide of Silver ¼ gr.; Powdered Ginger ½ gr.; Liquorice Root in Powder 2 grs.; Treacle *q. s.*: in one pill. *London.*

Pilula Argenti Oxidi et Opii.

Oxide of Silver ½ gr.; Opium ½ gr.; Mucilage *q. s. City Chest.*

Pilula Assafœtidæ c. Ferro.

Assafœtida 2 grs.; Sulphate of Iron 1 gr.; Extract of Gentian 1 gr. *Consumption.*

Comp. Assafœtida Pill 4 grs.; Dried Sulphate of Iron 1 gr.: in one pill. *London.*

Comp. Assafœtida Pill 4 grs.; Sulphate of Iron 1 gr.: in one pill. *City Chest.*

Pilula Assafœtidæ et Zinci.
Sulphate of Zinc 1 gr. ; Comp. Assafœtida Pill 4 grs. : for one pill. *City Chest.*

Pilula Atropiæ.
Atropia $\frac{1}{200}$ gr. ; Rectified Spirit $\frac{1}{20}$ min. ; Liquorice Root 1 gr. ; Treacle *q. s.* : in one pill. *St. Bartholomew's.*

Pilula Atropiæ et Morphiæ.
Sulphate of Atropine $\frac{1}{40}$ gr. ; Hydrochlorate of Morphia $\frac{1}{4}$ gr. ; Capsicum 1 gr. ; Pill of Aloes and Myrrh 3 grs. *City Chest.*

Pilula Atropiæ Sulphatis.
Sulphate of Atropine $\frac{1}{50}$, $\frac{1}{75}$, $\frac{1}{100}$ gr. : in one pill. *City Chest.*
Sulphate of Atropia $\frac{1}{75}$, $\frac{1}{50}$ gr. ; Confection of Roses *q. s. Gt. Northern.*

Pilula Belladonnæ.
Extract of Belladonna $\frac{1}{4}$ or $\frac{1}{2}$ gr. ; Extract of Gentian 1$\frac{1}{2}$ gr. ; Cinnamon Powder *q. s. Consumption.*
Extract of Belladonna $\frac{1}{4}$ gr. : in one pill. *City Chest.*
Extract of Belladonna $\frac{1}{4}$ gr. ; Extract of Gentian 4 grs. *Royal Chest.*
Extract of Belladonna $\frac{1}{3}$ gr. ; Liquorice Root 2 grs. ; Treacle *q. s.* : in one pill. *St. Bartholomew's.*

Pilula Belladonnæ c. Ipecacuanhâ.
Extract of Belladonna $\frac{1}{4}$ gr. ; Ipecacuanha $\frac{1}{2}$ gr. ; Extract of Taraxacum 3 grs. : in one pill. *Consumption.*

Pilula Belladonnæ et Zinci Sulph.
Extract of Belladonna $\frac{1}{8}$ gr. ; Sulphate of Zinc 1 gr. ; Sugar of Milk 1 gr. ; Treacle *q. s. London Ophthalmic.*

Pilula Bismuthi.
Subnitrate of Bismuth 5 grs. ; Mucilage *q. s.* : in one pill. *City Chest.*

Pilula Bismuthi et Conii.
Subnitrate of Bismuth 4 grs. ; Extract of Conium 2 grs. : in one pill. *City Chest.*

Pilula Bismuthi et Creasoti Composita.
Subnitrate of Bismuth 5 grs.; Opium in powder ¼ gr.; Creasote ½ min.; Glycerine of Tragacanth *q. s.*: in one pill. *London.*

Pilula Bismuthi et Hyoscyami.
Subnitrate of Bismuth 4 grs.; Extract of Henbane 2 grs.: in one pill. *City Chest.*

Pilula Calcii Sulphidi.
Sulphide of Calcium $\frac{1}{10}$ gr.; Excipient *q. s. Royal Free.*
Sulphide of Calcium ⅓ gr.; Compound Tragacanth Powder 1 gr.; water *q. s.* (varnished). *University.*
Sulphide of Calcium ½ gr.; Liquorice Root in powder 2 grs.; Treacle *q. s.*: in one pill (varnish). *London.*

Pilula Calcii Sulphidi Fort.
Sulphide of Calcium ¼ gr.; Excipient *q. s. Royal Free.*

Pilula Calomelanos. *See* Pil. Hydr. Subchlor.

Pilula Cambogiæ Aloes et Rhei.
Rhubarb ½ gr.; Aloes ¾ gr.; Gamboge ¼ gr.; Ginger 1 gr.; Treacle *q. s. Royal Chest.*

Pilula Cambogiæ Comp.
Camboge ½ gr.; Comp. Extract of Colocynth 2½ grs.; Extract of Henbane 2 grs.: in one pill. *Samaritan.*
Camboge 1 gr.; Socotrine Aloes 1 gr.; Jalap 3 grs.; Oil of Cloves ⅓ min.; water *q. s.*: in one pill. *London.*

Pilula Camphoræ et Hyoscyami.
Camphor 2 grs.; Extract of Henbane 1½ gr.; Mucilage *q. s.*: in one pill. *Samaritan.*
Camphor 2 grs.; Extract of Henbane 3 grs.: in one pill. *City Chest. Consumption. Fever. Guy's. Middlesex. Westminster.*
Camphor 2½ grs.; Extract of Henbane 2½ grs. *St. Bartholomew's. Westminster Ophthalmic.*
Camphor 3 grs.; Rectified Spirit 1 min.; Extract of Henbane 2 grs.: in one pill. *London.*

Pilula Camphoræ c. Opio.

Camphor 2 grs.; Opium ½ gr.; Treacle q. s.: in one pill. *St. Bartholomew's.*

Camphor 2 grs.; Opium ¼ gr.; Extract of Henbane 2 grs.; Rectified Spirit q. s. *Royal Free.*

Camphor 1 gr.; Opium 1 gr.; Extract of Hop sufficient: in one pill. *Consumption.*

Pilula Cannabis Indicæ.

Extract of Indian Hemp ¼ gr.; Extract of Gentian 2 grs.; Bread Mass 1 gr. *Royal Chest.*

Extract of Indian Hemp ¼ gr.; Sugar of Milk 2 grs.; Treacle q. s. *London Ophthalmic.*

Extract of Indian Hemp ½ gr.; Liquorice 1½ gr.; Mucilage q. s.: in one pill. Dose 1 or 2. *Consumption.*

Pilula Cathartica.

Calomel 8 grs.; Scammony 8 grs.; Colocynth Pulp 8 grs.; Gamboge 8 grs.; Barbadoes Aloes 12 grs.; Jalap 16 grs.; Croton Oil 1 drop; Oil of Peppermint 2 drops; Glycerine and Treacle q. s.: divide into 16 pills. Dose 1 or 2. *Royal Free.*

Socotrine Aloes 1 gr.; Jalap 3 grs.; Gamboge 1 gr.; Oil of Cloves ⅓ min.; water q. s.: in one pill. *Throat.*

Pilula Cathartica c. Hydrarg. Subchlor.

Socotrine Aloes 1 gr.; Jalap 3 grs.; Calomel ½ gr.; water q. s.: in one pill. *London.*

Pilula Cerii Oxalatis.

Oxalate of Cerium 3 grs.; Extract of Gentian q. s. *City Chest. London.*

Pilula Codeiæ.

Codeia ⅓ gr.; Liquorice Root in powder 2 grs.; Treacle q. s.: in one pill. *London.*

Pilula Colchici Acetici.

Acetic Extract of Colchicum 1½ gr.; Heavy Carbonate of Magnesia q. s.: in one pill. *St. Mary's.*

Pilula Colchici c. Belladonnâ.

Acetic Extract of Colchicum 1 gr.; Extract of Belladonna ½ gr; Extract of Chamomile 2 grs.: in one pill. *St. Thomas's.*

Pilula Colchici Co.
Acetic Extract of Colchicum 1 gr.; Dover's Powder 3 grs.: in one pill. *Middlesex.*
Acetic Extract of Colchicum 1 gr.; Compound Powder of Ipecacuanha 4 grs.: in one pill. *City Chest.*
Acetic Extract of Colchicum 1 gr.; Blue Pill 2 grs.; Gentian Root 2 grs. *Charing Cross.*
Extract of Colchicum ½ gr.; Calomel 1 gr.; Comp. Powder of Ipecacuanha 2½ grs. *University.*

Pilula Colchici c. Ipecac. Co.
Acetic Extract of Colchicum 1 gr.; Dover's Powder 4 grs.; Treacle *q. s.*: in one pill. *London.*

Pilula Colchici c. Opio.
Acetic Extract of Colchicum 2 grs.; Dover's Powder 5 grs.: in two pills. *St. George's.*

Pilula Colchici et Quiniæ.
Acetic Extract of Colchicum 1 gr.; Sulphate of Quinia 1 gr. *London Ophthalmic.*

Pilula Colocynthidis c. Assafœtidâ.
Comp. Colocynth Pill 2 grs.; Comp. Assafœtida Pill 3 grs.: in one pill. Dose 1 or 2. *Consumption.*

Pilula Colocynthidis c. Hydrargyro.
See Pil. Hydrarg. c. Colocynthide.

Pilula Colocynthidis c. Hydrarg. et Ipecac.
Pill of Colocynth and Mercury 4 grs.; Ipecacuanha 1 gr.: in one pill. *London.*

Pilula Colocynth. c. Hydr. Subchlor.
See Pil. Hyd. Subchlor. c. Colocynth.

Pilula Colocynthidis c. Hyoscyamo.
Comp. Colocynth Pill 3 grs.; Extract of Henbane 2 grs. *Westminster Ophthalmic.*

Pilula Colocynth. c. Ipecac. vel Pil. Aperiens.
Comp. Colocynth Pill 2 grs.; Blue Pill 1½ gr.; Ipecacuanha ⅓ gr.; Extract of Henbane 1 gr.: in one pill. *University.*

Pilula Colocynthidis c. Oleo Crotonis.
Comp. Colocynth Pill 4 grs.; Capsicum ¼ gr.; Oil of Croton ¹⁄₁₂ min.: in one pill. *Consumption*.
Comp. Colocynth Pill 4 grs.; Oil of Croton ½ min. *King's*.

Pilula Colocynthidis c. Opio.
Opium Pill (now Comp. Soap Pill) 5 grs.; Comp. Extract of Colocynth 5 grs.: in two pills: *St. George's*.

Pilula Colocynthidis c. Rheo.
Compound Colocynth Pill 2 grs.; Compound Rhubarb Pill 3 grs. *University*.
Compound Ext. of Colocynth 2½ grs.; Compound Rhubarb Pill 2½ grs. *Women*.

Pilula Colocynthidis Scammonii et Rhei.
Comp. Extract of Colocynth 3 grs.; Scammony 1 gr.; Extract of Rhubarb 1 gr.; Oil of Cinnamon ½ drop. *St. Thomas's*.

Pilula Conii et Acidi Benzoici.
Extract of Conium 2 grs.; Benzoic Acid 3 grs. *Guy's*.

Pilula Conii et Bismuth.
Extract of Conium 2⅓ grs.; Subnitrate of Bismuth 3⅓ grs.: in one pill. *Guy's*.

Pilula Conii Co.
Extract of Conium 300 grs.; Ipecacuanha 60 grs.; Treacle *q. s.*: divide into 78 pills. *St. George's*.

Pilula Conii et Hydrargyri.
Extract of Conium 4⅙ grs.; Blue Pill 1 gr. *Guy's*.

Pilula Conii et Ipecac. Comp.
Extract of Conium, Ipecacuanha, Extract of Henbane, Comp. Squill Pill, of each 1 gr. *Royal Chest*.

Pilula Conii c. Morphiâ.
Hydrochlorate of Morphia ⅙ gr.; Compound Pill of Conium 5 grs. *Charing Cross*.

Hydrochlorate of Morphia $\frac{1}{8}$ gr. ; Extract of Conium 4 grs. ; Ipecacuanha $\frac{1}{2}$ gr. : in one pill. *Consumption. Royal Chest. Gt. Northern.*
Hydrochlorate of Morphia $\frac{1}{8}$ gr. ; Extract of Conium 4 grs. *Westminster.*

Pilula Conii et Zinci.
Sulphate of Zinc 2 grs. ; Extract of Conium 3 grs. : in one pill. *City Chest.*

Pilula Creasoti.
Creasote 1 min. ; Crumb of Bread *q. s.* : for one pill. *City Chest. Gt. Northern. London* (varnish). *St. Bartholomew's.*
Creasote 1 min. ; Yellow Wax 3 grs. *Guy's.*

Pilula Creasoti c. Assafœtidâ.
Creasote 1 min. ; Comp. Assafœtida Pill 2 grs. ; Comp. Rhubarb Pill 2 grs. : in one pill. *St. Bartholomew's.*

Pilula Crotonis Olei.
Croton Oil $\frac{1}{4}$ min. ; Comp. Extract of Colocynth 5 grs. *British Skin.*
Croton Oil 1 min. ; Crumb of Bread *q. s.* : in one pill. *London.*
Croton Oil $\frac{1}{2}$ min. ; Hard Soap 2 grs. Comp. Extract of Colocynth $2\frac{1}{2}$ grs. *Charing Cross.*

Pilula Crotonis Co.
Croton Oil $\frac{1}{2}$ min. ; Comp. Pill of Colocynth 4 grs. : in one pill. *Westminster Ophthalmic.*

Pilula Croton-Chloral.
Croton-Chloral Hydrate 4 grs, ; Comp. Powder of Tragacanth 1 gr. ; water *q. s.* : in one pill. *London.*

Pilula Cupri Co.
Sulphate of Copper $\frac{1}{4}$ gr. ; Powdered Opium $\frac{1}{8}$ gr. : Extract of Gentian to 3 grs. : in one pill. *Middlesex.*
Sulphate of Copper $\frac{1}{4}$ gr. ; Opium $\frac{1}{4}$ gr. ; Confection of Roses *q. s. Fever.*

Pilula Cupri Sulphatis.
Sulphate of Copper $\frac{1}{4}$ gr. ; Opium in powder $\frac{1}{2}$ gr. ; Confection of Hips *q. s.* : for one pill. *Guy's.*

Pilula Cupri Sulphatis c. Opio.

Sulphate of Copper ¼ gr.; Opium ½ gr.; Extract of Gentian 2 grs.: in one pill. Dose 1 or 2 pills. *Consumption.*

Sulphate of Copper ¼ gr.; Extract of Opium ¼ gr.; Confection of Roses *q. s.*: in one pill. *City Chest. Gt. Northern.*

Sulphate of Copper ½ gr.; Opium ¼ gr.; Extract of Gentian *q. s. Royal Chest.*

Sulphate of Copper ½ gr.; Opium ¼ gr.; Liquorice Root in powder 2 grs.; Confection of Hips *q. s.*: in one pill. *London.*

Sulphate of Copper ½ gr.; Extract of Opium ½ gr.; Confection of Roses *q. s. Royal Free.*

Pilula Digitalis.

Digitalis ½ gr.; Sulphate of Iron ½ gr.; Capsicum ¼ gr.; Comp. Rhubarb Pill 1½ gr.; in one pill. *Samaritan.*

Pilula Digitalis Co.

Digitalis Powder ½ gr.; Squill 1 gr.; Blue Pill 3 grs.: in one pill. *St. George's.*

Digitalis ½ gr.; Squill 1 gr.; Blue Pill 1 gr.; Treacle *q. s.*: in one pill. *Middlesex.*

Pilula Digitalis c. Hydrargyro.

Digitalis Powder 1 gr.; Squill 2 grs.; Blue Pill 2 grs.: in one pill. *Consumption.*

Digitalis Powder ½ gr.; Squill 1 gr.; Blue Pill 1 gr.; Extract of Henbane 2 grs.: in one pill. *London.*

Pilula Digitalis Plumbea.

Acetate of Lead ⅙ gr.; distilled water 1 min.; dissolve and add Opium, in powder, 1/10 gr.; Digitalis Leaf ⅙ gr.; Liquorice Root 1 gr.; Glycerine of Tragacanth *q. s.*: in one pill. *Throat.*

Pilula Digitalis c. Scillâ.

Digitalis Powder ½ gr.; Compound Squill Pill 4½ grs.: in one pill. *Middlesex.*

Pilula Diuretica.

Blue Pill 1 gr.; Squill 1 gr.; Digitalis 1 gr.; Extract of Henbane 2 grs. *Gt. Northern.*

Pilula Diuretica c. Hydrargyro.
Blue Pill 1 gr.; Squill 1 gr.; Digitalis Powder ½ gr.; Ipecacuanha ½ gr.: in one pill. *London.*
Blue Pill 1 gr.; Squill 1 gr.; Digitalis Powder 1 gr.; Extract of Henbane 1⅔ gr. *Guy's.*

Pilula Doveri et Hydrargyri.
Dover's Powder 4 grs.; Blue Pill 1 gr.: in one pill. *City Chest.*

Pilula Doveri et Hydrargyri cum Cretâ.
Dover's Powder 2½ grs.; Grey Powder 2½ grs.; Mucilage *q. s.*: in one pill. *City Chest.*

Pilula Doveri et Hyoscyami.
Dover's Powder 2½ grs.; Extract of Henbane 2½ grs.: in one pill. *City Chest.*

Pilula Doveri et Hyoscyami et Scillæ.
Dover's Powder 2 grs.; Extract of Henbane 2 grs.; Squill 1 gr.: in one pill. *City Chest.*

Pilula Doveri et Hyoscyami et Zinci.
Dover's Powder 2 grs.; Extract of Henbane 2 grs.; Oxide of Zinc 1 gr.: in one pill. *City Chest.*

Pilula Doveri et Papaveris.
Dover's Powder 2½ grs.; Extract of Poppies 2½ grs.: in one pill. *City Chest.*

Pilula Doveri et Scillæ et Digitalis.
Dover's Powder 2 grs.; Squill 1 gr.; Digitalis 1 gr.; Confection of Roses *q. s.*: in one pill. *City Chest.*

Pilula Elaterii.
Elaterium 1/12 gr.; Extract of Henbane 4 grs.: in one pill. *St. Mary's.*
Elaterium ⅛ gr.; Extract of Belladonna ¼ gr.; Capsicum 1 gr.; Extract of Jalap to 5 grs. *Charing Cross.*
Elaterium ¼ gr.; Calomel ½ gr.; Capsicum 1 gr.; Extract of Poppies 1 gr.: in one pill. *Samaritan.*

Pilula Elaterii Co.
Elaterium ⅙ gr.; Extract of Socotrine Aloes 1 gr.; Extract of Henbane 2 grs.: in one pill. *Middlesex.*

Elaterium $\frac{1}{12}$ gr. ; Comp. Gamboge Pill 2½ grs. ; Compound Colocynth Pill 2½ grs. ; Oil of Cassia ½ min : in one pill. *City Chest.*
Elaterium $\frac{1}{6}$ gr. ; Pill of Colocynth and Henbane 5 grs. : in one pill. *London.*
Elaterium ¼ gr. ; Comp. Extract of Colocynth 2 grs. ; Calomel 1½ gr. ; Capsicum ½ gr. ; Treacle *q. s.* : in one pill. *St. Thomas's. St. Bartholomew's.*

Pilula Ergotæ et Opii.
Extract of Ergot 1 gr. ; Tannic Acid 2 grs. ; Extract of Opium ¼ gr. : in one pill. *City Chest. Gt. Northern.*

Pilula Expectorans.
Ipecacuanha ½ gr. ; Guaiacum 1 gr. ; Opium ½ gr. : Comp. Squill Pill 2 grs. : in one pill. *City Chest. Gt. Northern.*
Ipecacuanha ¾ gr. ; Squill 1½ gr. ; Opium ¼ gr. ; Hard Soap 1 gr. ; Glycerine *q. s. Royal Free.*
Ipecacuanha ½ gr. ; Extract of Conium 2 grs. ; Powder of Squill 1½ gr. ; Tartar Emetic $\frac{1}{24}$ gr. : in one pill. *Samaritan.*

Pilula Ferri. *See* Pilula Ferri Redacti.

Pilula Ferri et Aloes.
Sulphate of Iron 1 gr. ; Aloes and Myrrh Pill 4 grs. : in one pill. *St. George's.*

Pilula Ferri Aperiens.
Dried Sulphate of Iron 1 gr. ; Socotrine Aloes 1 gr. ; Extract of Nux Vomica ¼ gr. ; Liquorice Root in powder 2 grs. ; Syrup *q. s.* : in one pill. *London.*

Pilula Ferri Arseniatis.
Arseniate of Iron $\frac{1}{16}$ gr. ; Extract of Gentian *q. s.* : in one pill. Dose 1 or 2 pills. *Consumption.*
Arseniate of Iron $\frac{1}{15}$ gr. ; Extract of Gentian 3 grs. *Charing Cross.*

Pilula Ferri Co.
Sulphate of Iron 1 gr. ; Carbonate of Soda 1 gr. ; Treacle 1 gr. ; Myrrh 2 grs. : in one pill. *City Chest.*

Dried Sulphate of Iron 2½ grs. ; Carbonate of Potash 2½ grs. ; Tragacanth Powder ⅛ gr. ; Syrup *q. s.*: in one pill. *London.*

Pilula Ferri Fœtida.

Saccharated Carbonate of Iron 3 grs. ; Comp. Assafœtida Pill 2 grs. ; Treacle *q. s.* : in one pill. *St. Bartholomew's.*

Pilula Ferri et Gentianæ.

Sulphate of Iron 1½ gr. ; Extract of Gentian 1½ gr. ; Ginger 1½ gr. : in one pill. *King's.*

Pilula Ferri et Potassæ Carbonatis.

Sulphate of Iron 2½ grs. ; Carbonate of Potash 2½ grs. ; Mucilage of Tragacanth *q. s.* : in one pill. *City Chest.*

Pilula Ferri et Quiniæ.

Sulphate of Iron 1 gr. ; Sulphate of Quinia 1 gr. ; Extract of Chamomile 1 gr. *Royal Chest.*

Sulphate of Iron 1 gr. ; Sulphate of Quinia 1 gr. ; Rhubarb ¾ gr. ; Ginger ½ gr. : in one pill. *Samaritan.*

Pilula Ferri Quiniæ et Aloes.

Sulphate of Iron 1 gr. ; Sulphate of Quinia 1 gr. ; Extract of Aloes ⅓ gr. ; Extract of Chamomile 1 gr. *Royal Chest.*

Pilula Ferri et Quiniæ Co.

Sulphate of Iron ½ gr. ; Sulphate of Quinia ½ gr. ; Ipecacuanha 1 gr. ; Comp. Squill Pill 1 gr. ; Extract of Hemlock 1 gr. *Royal Chest.*

Pilula Ferri Redacti.

Reduced Iron 3 grs. ; Liquorice Root 1 gr. ; Mucilage of Tragacanth *q. s.* : in one pill. *St. Bartholomew's.*

Reduced Iron 3 grs. ; Liquorice Root in powder 2 grs. ; Balsam of Peru *q. s.* : in one pill. *London.*

Reduced Iron 4 grs. ; Confection of Roses *q. s. Women.*

Pilula Ferri Redacti et Hydrarg. c. Cretâ.

Reduced Iron 2 grs. ; Grey Powder 1 gr. ; Liquorice Root, in powder, 2 grs. ; Treacle *q. s.* : in one pill. *London.*

Pilula Ferri Redacti et Quiniæ.
Reduced Iron 3 grs.; Sulphate of Quinine 1 gr.; Liquorice Root in powder 1 gr.; Balsam of Peru *q. s.*: in one pill. *London.*

Pilula Ferri c. Strychniâ.
Dried Sulphate of Iron 2 grs.; Strychnia $\frac{1}{16}$ gr.; Extract of Henbane 3 grs.: in one pill. *London.*

Pilula Ferri Sulphatis.
Sulphate of Iron 1 gr.; Ginger 1 gr.; Extract of Gentian 3 grs. *Gt. Northern.*
Dried Sulphate of Iron 1 gr.; Ginger 1 gr.; Extract of Gentian 3 grs.: in one pill. *St. Bartholomew's.*
Dried Sulphate of Iron 2 grs.; Liquorice Root in powder, 2 grs.; Treacle *q. s.*: in one pill. *London.*
Dried Sulphate of Iron 5 grs.; Syrup 1 min. *University.*

Pilula Ferri Sulphatis Exsiccati.
Dried Sulphate of Iron 4 grs.; Bread Crumb and Treacle of each *q. s.*: in one pill. *City Chest.*
Dried Sulphate of Iron 2 grs.; Treacle *q. s. Middlesex.*

Pilula Ferri c. Tiglio.
Sulphate of Iron 1 gr.; Croton Oil $\frac{1}{12}$ min.; Comp. Rhubarb Pill $1\frac{2}{3}$ gr.: in one pill. *Samaritan.*

Pilula Ferri et Valerian.
Sulphate of Iron 15 grs.; Strychnia 1 gr.; Extract of Rhubarb 40 grs.; Valerianate of Quinia 15 grs. Mix and divide into 30 pills. *Samaritan.*

Pilula Ferri Valerianatis.
Valerianate of Iron 2 grs.; Acacia 1 gr.; Extract of Henbane 1 gr. *Women.*

Pilula Ferri Valerianatis et Aloes.
Valerianate of Iron 1 gr.; Extract of Aloes $\frac{1}{2}$ gr.; Extract of Gentian 2 grs.: in one pill. *City Chest.*

Pilula Ferri et Zinci.
Sulphate of Iron 1 gr.; Sulphate of Zinc 1 gr.; Ipecacuanha 1 gr.; Extract of Conium 2 grs.: in one pill. *City Chest.*

Pilula Ferri et Zinci Valerianatis.
Valerianate of Iron 1 gr.; Valerianate of Zinc 2 grs.; Extract of Gentian 2 grs.: in one pill. *London.*

Pilula Fœtida c. Ferro. See Pilula Ferri Fœtida.

Pilula Galbani Co.
Ph. Lond. *St. George's.*

Pilula Galbani et Rhei.
Comp. Galbanum Pill $2\frac{1}{2}$ grs.; Comp. Rhubarb Pill $2\frac{1}{2}$ grs.: in one pill. *St. George's.*

Pilula Gentianæ et Ferri.
Extract of Gentian 3 grs.; Sulphate of Iron 1 gr.; *Guy's.*

Pilula Gentianæ et Zinci.
Extract of Gentian 3 grs.; Sulphate of Zinc 1 gr.; Calumba in powder *q. s.*: for one pill. *Guy's.*

Pilula Hydrargyri Bichloridi c. Opio.
Perchloride of Mercury $\frac{1}{12}$ gr.; Powder of Opium $\frac{1}{3}$ gr.; Confection of Roses *q. s.*: in one pill. *Samaritan.*

Pilula Hydrargyri Bromidi.
Bromide of Mercury $\frac{1}{2}$ gr.; Confection of Roses *q. s.* *Women.*

Pilula Hydrargyri c. Colocynthide.
Blue Pill 1 gr.; Extract of Colocynth 2 grs.; Extract of Henbane 2 grs. Dose 1 to 2 pills. *St. Mary's.*

Blue Pill 1 gr.; Comp. Extract of Colocynth $2\frac{1}{2}$ grs.: in one pill. *St. Bartholomew's.*

Blue Pill 1 gr.; Comp. Extract of Colocynth 4 grs.: in one pill. *Samaritan.*

Blue Pill $1\frac{1}{2}$ gr.; Colocynth and Hyoscyamus Pill $3\frac{1}{2}$ grs.; in one pill. Dose 1 or 2 pills. *Consumption.*

Blue Pill 2 grs.; Comp. Pill of Colocynth 3 grs.:
in one pill. *London. London Ophthalmic. University. Westminster Ophthalmic.*
Blue Pill 2 grs.; Colocynth Pill with Henbane 3
grs. *British Skin. Women.*
Blue Pill 2 grs.; Comp. Extract of Colocynth 2 grs.;
Ipecacuanha ⅓ gr.; Extract of Hyoscyamus
1 gr. *Westminster.*
Blue Pill 2½ grs.; Comp. Pill of Colocynth 2½ grs.;
Middlesex. Royal Free. St. George's.

Pilula Hydrarg. c. Coloc. et Hyoscyam.
Blue Pill 1 gr.; Comp. Extract of Colocynth 2½
grs.; Extract of Henbane 1 gr. *St. Bartholomew's.*

Pilula Hydrargyri Co.
Mercurial Pill 2 grs.; Powder of Ipecacuanha and
Opium 3 grs. *Fever.*
Mercurial Pill 1½ gr.; Ipecacuanha ¼ gr.; Acetic
Extract of Colchicum ¼ gr.; Pill of Colocynth
and Henbane 3 grs.: in one pill. *London.*

Pilula Hydrargyri c. Cretâ.
Mercury with Chalk 1 gr.; Compound Powder of
Ipecacuanha 2 grs.; Glycerine of Tragacanth
q. s.: in one pill. *Throat.*

Pilula Hydrargyri c. Cretâ Co.
Grey Powder 2½ grs.; Dover's Powder 2½ grs.;
Treacle *q. s.*: in one pill. *St. Bartholomew's. Westminster Ophthalmic.*

Pilula Hydrarg. c. Cretâ et Pulv. Ipecac. Co.
(*See* also Pil. Doveri et Hydrarg. c. Cretâ.)
Grey Powder 2 grs.; Dover's Powder 1 gr.; Treacle
q. s.: in one pill. *London.*
Grey Powder 2 grs.; Dover's Powder 3 grs.; Mucilage *q. s. London Ophthalmic.*

Pilula Hydrarg. c. Cretâ et Ipecac.
Grey Powder 2½ grs.; Dover's Powder 2½ grs.;
Confection of Hips *q. s.*: for one pill. *University.*

Pilula Hydrarg. c. Cretâ et Quinia.
Grey Powder 3 grs.; Sulphate of Quinine 1 gr.;
Treacle *q. s.*: in one pill. *St. Bartholomew's.*

Pilula Hydrarg. c. Cretâ et Rheo.
Grey Powder 2½ grs.; Rhubarb 2½ grs.; Treacle q. s.: in one pill. *St. Bartholomew's.*

Pilula Hydrargyri Cyanidi.
Cyanide of Mercury 1/10 gr.; Sugar of Milk ¾ gr.; Glycerine of Tragacanth q. s.: for one varnished pill. *Throat.*

Pilula Hydrargyri Diuretica.
Blue Pill 1 gr.; Digitalis 1 gr.; Squill 2 grs.: in one pill. Dose 1 to 2 pills. *St. Mary's.*
Blue Pill 2 grs.; Squill 2 grs.; Digitalis 1 gr. *Charing Cross.*

Pilula Hydrargyri c. Hyoscyamo.
Blue Pill 1½ gr.; Extract of Henbane 2 grs.; Extract of Socotrine Aloes 1½ gr.: in one pill. *Middlesex.*
Blue Pill 3 grs.; Extract of Henbane 2 grs. *Gt. Northern.*
Blue Pill 2½ grs.; Extract of Henbane 2½ grs.: in one pill. *London.*

Pilula Hydrargyri Iodidi Viridis.
Green Iodide of Mercury ½ gr.; Powdered Opium ¼ gr.; Extract of Gentian 2 grs. *British Skin.*
Green Iodide of Mercury ½ gr.; Powdered Opium ½ gr.; Glycerine of Tragacanth q. s.: in one pill. *Throat.*

Pilula Hydrarg. Iodid. Virid. cum Opio.
Green Iodide of Mercury ½ gr.; Powdered Opium ⅛ gr.; Treacle q. s.: in one pill. *St. Bartholomew's.*

Pilula Hydrargyri c. Opio.
Blue Pill 1½ gr.; Opium ½ gr.: in one pill. *Throat.*
Blue Pill 2 grs.; Opium ⅕ gr. *University,* No. 1.
Blue Pill 3 grs.; Opium ¼ gr.: in one pill. *King's. Royal Free. University,* No. 2.
Blue Pill 4 grs.; Opium ¼ gr.: in one pill. *Guy's. London.*

Blue Pill 4½ grs.; Opium ½ gr.: in one pill. *St. George's.*
Blue Pill 5 grs.; Opium ½ gr. *London Ophthalmic. University,* No. 3.
Blue Pill 5 grs.; Opium ⅓ gr.: in one pill. *Westminster Ophthalmic.*

Pilula Hydrarg. Perchlor. c. Belladonnâ.

Perchloride of Mercury $\frac{1}{16}$ gr.; Sulphate of Quinine 1 gr.; Extract of Belladonna ½ gr.; Extract of Gentian 3 grs.: in one pill. *London.*

Pilula Hydrarg. c. Pulv. Ipecac. Co.

Blue Pill 2 grs.; Dover's Powder 3 grs.: in one pill. *Consumption.*
Blue Pill 2½ grs.; Dover's Powder 2½ grs.: in one pill. *London.*

Pilula Hydrarg. c. Quiniâ et Ferro.

Blue Pill 1 gr.; Sulphate of Quinia 1 gr.; Sulphate of Iron 2 grs.; Treacle *q. s.*: in one pill. *Westminster Ophthalmic.*

Pilula Hydrargyi c. Rheo.

Blue Pill 1½ gr.; Comp. Rhubarb Pill 3 grs.; Ipecacuanha ½ gr. *Charing Cross.*
Blue Pill 1 gr.; Comp. Rhubarb Pill 3½ grs. *Royal Chest.*
Comp. Rhubarb Pill 4 grs.; Mercurial Pill 1 gr. *Fever.*
Blue Pill 1½ gr.; Rhubarb 3 grs.; Ipecacuanha ½ gr.; Treacle *q. s.*: in one pill. *Middlesex.*
Blue Pill 1½ gr.; Comp. Rhubarb Pill 3 grs.: in one pill. Dose 1 or 2 pills. *Consumption.*
Blue Pill 2 grs.; Compound Rhubarb Pill 3 grs. *City Chest. St. Bartholomew's.*
Blue Pill 2½ grs.; Comp. Rhubarb Pill 2½ grs. *Guy's. London. London Ophthalmic. Royal Free. St. George's. St. Thomas's. Westminster Ophthalmic.*

Pilula Hydrargyri c. Scillâ.

Blue Pill 1 gr.; Comp. Squill Pill 2 grs.; Extract of Henbane 2 grs. *Charing Cross.*

Blue Pill 1 gr.; Comp. Squill Pill 4 grs. *Consumption.*
Blue Pill 2 grs.; Squill 2 grs.: in one pill. *City Chest.*
Blue Pill $2\frac{1}{2}$ grs.; Compound Squill Pill $2\frac{1}{2}$ grs.: in one pill. *London.*
Blue Pill 3 grs.; Squill $1\frac{1}{2}$ gr.: in one pill. *St. Bartholomew's.*
Comp. Squill Pill 4 grs.; Grey Oxide of Mercury $\frac{1}{3}$ gr.: in one pill. *Guy's.*

Pilula Hydrarg. Scillæ et Digitalis.
Blue Pill $1\frac{1}{2}$ gr.; Squill $1\frac{1}{2}$ gr.; Digitalis $\frac{1}{2}$ gr. *Royal Chest. Westminster.*

Pilula Hydrargyri Subchloridi.
Calomel $\frac{1}{2}$ to 5 grs.; Liquorice Root, in powder, and Treacle *q. s.*: in one pill. *London.*

Pilula Hydrarg. Subchlor. c. Colocynth.
Calomel $\frac{1}{2}$ gr.; Ipecacuanha $\frac{1}{2}$ gr.; Comp. Pill of Colocynth 4 grs. *Royal Chest.*
Calomel 1 gr.; Comp. Extract of Colocynth 3 grs.; Extract of Henbane 1 gr. *London Ophthalmic.*
Calomel 1 gr.; Comp. Extract of Colocynth $3\frac{1}{2}$ grs.; Ipecacuanha $\frac{1}{2}$ gr.: in one pill. *Middlesex.*
Calomel 1 gr.; Comp. Colocynth Pill 3 grs. *University.*
Calomel 1 gr.; Comp. Colocynth Pill 4 grs.; in one pill. *Guy's. London. Westminster Ophthalmic. Women.*
Calomel 1 gr.; Comp. Colocynth Pill 4 grs; Oil of Caraway $\frac{1}{4}$ gr. *King's. Royal Free.*
Calomel 1 gr.; Comp. Extract of Colocynth 4 grs.: in one pill. *St. Bartholomew's. St. Thomas's.*
Calomel $1\frac{1}{2}$ gr.; Comp. Colocynth Pill 3 grs.; Capsicum $\frac{1}{6}$ gr.: for one pill. Dose 1 or 2 pills. *Consumption.*
Comp. Extract of Colocynth $3\frac{1}{2}$ grs.; Calomel $1\frac{1}{2}$ grs.: in one pill. *St. George's.*
Calomel $1\frac{1}{2}$ gr.; Comp. Colocynth Pill $2\frac{1}{2}$ grs.; Extract of Henbane $\frac{1}{2}$ gr.: in one pill. *Samaritan.*
Calomel 2 grs.; Comp. Extract of Colocynth 3 grs. *Charing Cross. St. Mary's. Westminster.*
Calomel 2 grs.; Comp. Colocynth and Hyoscyamus Pill 3 grs. *Gt. Northern.*

Pilula Hydrarg. Subchlor. c. Coloc. et Hyosc.

Calomel 1 gr.; Comp. Extract of Colocynth 3 grs.; Extract of Hyoscyamus 1 gr. *St. Bartholomew's.*

Pilula Hydrarg. Subchlor. c. Jalapâ.

Calomel 1 gr.; Jalap 3 grs.; Treacle *q. s.*: in one pill. *St. Bartholomew's.*

Pilula Hydrarg. Subchlor. c. Opio.

Calomel ½ to 2 grs.; Opium in powder ¼ to 1 gr.; Liquorice Root in powder 2 grs.; Treacle *q. s.*: in one pill. *London.*
Calomel 1 gr.; Opium ¼ gr.; Comp. Powder of Tragacanth 1 gr.; Glycerine of Tragacanth *q. s. University.*
Calomel 1 gr.; Opium ¼ gr.; Wheaten Flour 1 gr.; Confection of Roses *q. s. London Ophthalmic.*
Calomel 1 gr.; Opium ¼ gr.; Treacle *q. s. Gt. Northern.*
Calomel 1 gr.; Opium ¼ gr.; Confection of Hips *q. s. King's. Royal Free.*
Calomel 1 gr.; Opium ¼ gr.; Confection of Roses *q. s. Guy's.*
Calomel 1 gr.; Opium ⅓ gr.; Confection of Roses *q. s.*: in one pill. *Westminster Ophthalmic.* No. 1.
Calomel 2 grs.; Extract of Opium ¼ gr.; Treacle *q. s.*: in one pill. *City Chest.*
Calomel 2 grs.; Opium ⅓ gr.; Confection of Roses *q. s. Fever.*
Calomel 2 grs.; Opium ½ gr.; Extract of Liquorice 2½ grs.: in one pill. *St. Thomas's.*
Calomel 2 grs.; Opium 1 gr.; Confection of Roses *q. s. Westminster Ophthalmic,* No. 2.

Pilula Hydrarg. Subchlor. c. Pulv. Ipecac. Co.

Calomel 1 gr.; Dover's Powder 4 grs.; Mucilage *q. s. London Ophthalmic.*

Pilula Hydrarg. Subchlor. c. Rheo.

Calomel 1 gr.; Rhubarb 4 grs.; Treacle *q. s.*: in one pill. *St. Bartholomew's.*
Calomel 1 gr.; Comp. Rhubarb Pill 2 grs.; Extract of Hyoscyamus 1 gr.: in one pill. *Throat.*

Pilula Hydrarg. Subchlor. c. Scammonio.
Calomel 1 gr.; Scammony 3 grs.; Treacle *q. s.*: in one pill. *St. Bartholomew's.*

Pilula Hydrargyri Suboxidi et Digitalis et Scillæ.
Suboxide of Mercury 1 gr.; Squill 1 gr.; Digitalis 1 gr.; Extract of Henbane 2 grs.: in one pill. *City Chest.*

Pilula Hyoscyami.
Extract of Henbane 4 grs.: in one pill. *City Chest.*
Extract of Henbane 5 grs.: in one pill. *London.*

Pilula Hyoscyami c. Camphorâ.
See Pilula Camphoræ c. Hyoscyamo.

Pilula Hyoscyami c. Conio et Ipecac.
Extract of Henbane 2 grs.; Extract of Conium 2 grs.; Ipecacuanha 1 gr.: in one pill. *London.*

Pilula Hyoscyami et Morphia.
Extract of Henbane 3 grs.; Hydrochlorate of Morphia ⅓ gr. *Guy's.*

Pilula Hyoscyami et Pulv. Ipecac. Co.
Extract of Henbane 2 grs.; Dover's Powder 2 grs.: in one pill. *St. Thomas's.*

Pilula Hyoscyami et Scillæ.
Extract of Henbane 2 grs.; Comp. Squill Pill 2 grs.; Ipecacuanha 1 gr.: in one pill. *London.*

Pilula Hyoscyami et Stramonii.
Extract of Henbane 4 grs.; Extract of Stramonium ½ gr.: in one pill. *City Chest.*

Pilula Iodoformi.
Iodoform 2 grs.; Sugar of Milk 1 gr.; Glycerine of Tragacanth *q. s.*: in one pill. *Throat.*

Pilula Ipecacuanhæ Co.
Dover's Powder 5 grs.; Treacle *q. s.*: in one pill. *Guy's. London.*

Pilula Ipecacuanhæ Co. c. Hydr.
See Pilula Hydr. c. Pulv. Ipecac. Co.

Pilula Ipecacuanhæ Co. c. Scillâ.
Powder of Ipecacuanha and Opium 3 grs.; Squill ½ gr.; Extract of Conium 1½ gr. *Fever.*

Pilula Ipecacuanhæ Opiata c. Scillâ.
Ipecacuanha ½ gr.; Comp. Squill Pill 1 gr.; Dover's Powder 2 grs.; Extract of Aloes ⅙ gr.: in one pill. *Royal Chest.*

Pilula Ipecacuanhæ c. Opio.
Ipecacuanha ¼ gr.; Opium ½ gr.; Confection of Roses 2 grs. *Consumption.*

Pilula Ipecacuanhæ et Rhei.
Ipecacuanha 1 gr.; Comp. Rhubarb Pill 4 grs. *Westminster.*

Pilula Lactucæ c. Hyoscyamo.
Extract of Lettuce 2 grs.; Extract of Henbane 2 grs.: in one pill. Dose 1 or 2. *Consumption.*

Pilula Morphiæ.
Hydrochlorate of Morphia ¼, ⅓, ½, or 1 gr.; Liquorice Root in powder 3 grs.; Treacle *q. s.*: in one pill. *London.*
Hydrochlorate of Morphia ¼, ⅓, or ½ gr.: in one pill. *Consumption.*

Pilula Morphiæ Acet.
Acetate of Morphia from 1/12 to ½ gr.; Confection of Roses *q. s.*: in one pill. *St. Mary's.*

Pilula Morphiæ et Glycyrrhizæ.
Hydrochlorate of Morphia 3/16 gr.; Extract of Liquorice 3 grs.; Comp. Tragacanth Powder 5 grs. *Royal Chest.*

Pilula Morphiæ Hydrochloratis.
Hydrochlorate of Morphia ⅛ to 1 gr. in each pill. *Royal Chest.*
Hydrochlorate of Morphia from 1/12 to ½ gr.; Confection of Roses *q. s.*: in one pill. *St. Mary's.*

Pilula Morphiæ c. Hyoscyamo.
Hydrochlorate of Morphia ¼ gr.; Extract of Henbane 3 grs.: in one pill. *City Chest. St. Thomas's.*

Pilula Morphiæ et Hyoscyami et Ipecacuanhæ.
Hydrochlorate of Morphia ¼ gr.; Ipecacuanha 1 gr.; Extract of Henbane 3 grs.; in one pill. *City Chest.*

Pilula Morphiæ et Hyoscyam. et Stramon.
Hydrochlorate of Morphia ¼ gr.; Extract of Stramonium ½ gr.; Extract of Henbane 3 grs.; in one pill. *City Chest.*

Pilula Morphiæ et Hyoscyam. et Zinci.
Hydrochlorate of Morphia ¼ gr.; Oxide of Zinc 2 grs.; Extract of Henbane 3 grs.; in one pill. *City Chest.*

Pilula Morphiæ c. Lupulo.
Hydrochlorate of Morphia 1/12 gr.; Ipecacuanha ¼ gr.; Extract of Hops 2 grs.; in one pill. Dose 1 or 2 pills. *Consumption.*

Pilula Nucis Vomicæ.
Extract of Nux Vomica ¼ gr.; Extract of Gentian 2 grs.; in one pill. *Consumption.*
Extract of Nux Vomica ¼ gr.; Compound Extract of Colocynth 3 grs.; Hard Soap 1¾ gr. *Charing Cross.*
Extract of Nux Vomica ½ to 1 gr.; Extract of Gentian 2 grs.; Liquorice Root in powder 2 grs.; in one pill. *London.*
Extract of Nux Vomica ½ gr.; Capsicum ¼ gr.; Comp. Rhubarb Pill 3 grs.; in one pill. *Samaritan.*
Extract of Nux Vomica ½ gr.; Powdered Capsicum ¼ gr.; Comp. Rhubarb Pill 2 grs. *Guy's.*

Pilula Nucis Vomicæ cum Belladonnâ.
Extract of Nux Vomica ½ gr.; Extract of Belladonna ⅓ gr.; Compound Rhubarb Pill 2½ grs. *Royal Free.*

Pilula Nucis Vomicæ Cathartica.

Extract of Nux Vomica ¼ gr.; Pill of Colocynth and Henbane 2 grs.; Comp. Rhubarb Pill 1 gr.; Liquorice Powder 2 grs.; in one pill. *London.*

Extract of Nux Vomica ¼ gr.; Comp. Extract of Colocynth 1¼ gr.; Extract of Henbane 1¼ gr.; Comp. Rhubarb Pill 1¼ gr.; in one pill. *Throat.*

Pilula Nucis Vomicæ et Ferri.

Extract of Nux Vomica ½ gr.; Dried Sulphate of Iron 1 gr.; Extract of Barbadoes Aloes 1 gr. *Guy's.*

Pilula Nucis Vomicæ et Rhei.

Extract of Nux Vomica ¼ gr.; Extract of Henbane 1 gr.; Comp. Rhubarb Pill 4 grs.; in one pill. *City Chest.*

Pilula Opii.

Powdered Opium ¼ to 1 gr. in each. *Royal Chest.*

Pilula Opii et Conii.

Extract of Conium 3 grs.; Compound Soap Pill 2 grs. *British Skin.*

Pilula Pepsinæ.

Pepsine 5 grs.; Glycerine *q. s.*: in one pill. *Varnish. London.*

Pilula Pepsinæ et Quiniæ.

Pepsine 3 grs.; Sulphate of Quinia ¾ gr.; Extract of Belladonna ¼ gr.; Glycerine *q. s.*: in one pill. *City Chest.*

Pilula Picis.

Tar 2 grs.; Magnesia *q. s. Consumption.*
Tar 2 grs.; Lycopodium 1 gr.; Liquorice Powder *q. s.*: in one pill. *City Chest.*

Pilula Picis Co.

Tar 2½ grs.; Liquorice Powder 2½ grs.; in one pill. *St. Thomas's.*

Pilula Plumbi c. Opio.

Acetate of Lead 2 grs.; Opium $\frac{1}{6}$ gr.; Liquorice Powder 2 grs.; Treacle *q. s.*: in one pill. *London.*

Pilula Podophylli.

Resin of Podophyllum $\frac{1}{8}$ gr.; Extract of Hyoscyamus 2 grs.; Soap *q. s.* *St. Mary's.*

Resin of Podophyllum $\frac{1}{8}$ gr.; Extract of Henbane $1\frac{1}{2}$ gr.; Comp. Rhubarb Pill $1\frac{1}{2}$ gr.; Capsicum $\frac{1}{2}$ gr.: in one pill. *Throat.*

Resin of Podophyllum $\frac{1}{4}$ gr.; Extract of Hyoscyamus sufficient. *Westminster.*

Resin of Podophyllum $\frac{1}{4}$ to 1 gr.; Extract of Hyoscyamus 2 grs.; in one pill. *London.*

Resin of Podophyllum $\frac{1}{4}$ gr.; Barbadoes Aloes 1 gr.; Capsicum $\frac{1}{2}$ gr.; Extract of Belladonna $\frac{1}{4}$ gr.; Glycerine of Tragacanth *q. s.* *University.*

Resin of Podophyllum $\frac{1}{3}$ gr.; Capsicum 1 gr.; Extract of Belladonna $\frac{1}{4}$ gr.; Comp. Extract of Colocynth to 5 grs. *Charing Cross.*

Pilula Podophylli Co.

Resin of Podophyllum $\frac{1}{6}$ gr.; Extract of Henbane 2 grs.; Comp. Powder of Cinnamon *q. s.* *Women.*

Resin of Podophyllum $\frac{1}{3}$ gr.; Extract of Barbadoes Aloes 2 grs.; Extract of Belladonna $\frac{1}{4}$ gr.; Hard Soap 1 gr. *Gt. Northern.*

Pilula Podophylli et Creasoti.

Resin of Podophyllum $\frac{1}{8}$ gr.; Creasote $\frac{1}{2}$ min.; Extract of Henbane $2\frac{1}{2}$ gr.; Comp. Rhubarb Pill $2\frac{1}{2}$ grs.; in one pill. *City Chest.*

Pilula Podophylli et Extracti Colocynthidis.

Resin of Podophyllum $\frac{1}{4}$ to $\frac{1}{2}$ gr.; Comp. Extract of Colocynth $2\frac{1}{2}$ to 5 grs.: in one pill. *City Chest.*

Pilula Podophylli c. Hydrargyro.

Resin of Podophyllum $\frac{1}{2}$ gr.; Blue Pill 1 gr.; Extract of Henbane 2 grs. *Guy's.*

Pilula Podophylli et Hyoscyami.

Resin of Podophyllum $\frac{1}{4}$, $\frac{1}{3}$, or $\frac{1}{2}$ gr.; Extract of Henbane 2 grs.; in one pill. *City Chest.*

Pilula Podophylli cum Nuce Vomicâ.
Resin of Podophyllum ⅙ gr. ; Extract of Nux Vomica ¼ gr.; Extract of Gentian 2 grs. *British Skin.*

Pilula Podophylli et Rhei.
Resin of Podophyllum ½ gr. ; Comp. Rhubarb Pill 4 grs.: in one pill. *City Chest.*

Pilula Podophylli c. Rheo.
Resin of Podophyllum ⅓ gr.; Comp. Rhubarb Pill 4 grs. *Royal Chest.*

Pilula Purgans.
Extract of Barbadoes Aloes 1½ gr. ; Comp. Extract of Colocynth 2 grs. ; Extract of Jalap 1½ gr. ; Caraway Oil *q. s.* *Charing Cross.*

Pilula Quiniæ.
Sulphate of Quinia 1, 2, 2½ or 3 grs. ; Liquorice Powder and Treacle *q. s.* ; in one pill. *London.*
Sulphate of Quinia 1 to 2 grs. ; Extract of Gentian *q. s.* : in one pill. Dose 1 or 2. *Consumption.*

Pilula Quiniæ c. Digitale.
Sulphate of Quinia 2 grs. ; Digitalis ½ gr. ; Extract of Hops sufficient. *Consumption.*

Pilula Quiniæ, Digitalis, et Opii.
Sulphate of Quinia 2 grs . ; Digitalis 1 gr. ; Opium ¼ gr. ; Confection of Roses *q. s.* *City Chest.*

Pilula Quiniæ c. Ferro.
Sulphate of Quinia 1 gr.; Sulphate of Iron 1 gr. ; Extract of Hemlock 3 grs. *Fever.*
Sulphate of Quinia 1 gr. ; Dried Sulphate of Iron 1 gr. ; Liquorice Powder 2 grs. ; Treacle *q. s.*: in one pill. *London.*
Sulphate of Quinia 1 gr. ; Sulphate of Iron 1 gr.: Confection of Roses *q. s.*: in one pill. *Middlesex.*
Sulphate of Quinia 1 gr. ; Sulphate of Iron 1 gr.; Extract of Gentian 3 grs.: in one pill. *Consumption.*

Sulphate of Quinia 2 grs. ; Sulphate of Iron 2 grs.;
Extract of Gentian 1 gr. : in one pill. *St.
Thomas's.*
Sulphate of Quinia 2 grs. ; Sulphate of Iron 1 gr.;
Treacle *q. s.*: in one pill. *City Chest.*

Pilula Quiniæ c. Opio.
Sulphate of Quinine 1 gr. ; Powdered Opium ¼ gr. ;
Confection of Roses *q. s. Middlesex.*
Sulphate of Quinine 1, 2, 2½ or 3 grs. ; Opium in
powder ¼, ⅓, ½, or 1 gr. ; Glycerine of Tragacanth *q. s.*: in one pill. *London.*

Pilula Rhei et Creasoti.
Powdered Rhubarb 3 grs. ; Creasote ½ min. ; Extract
of Chamomile *q. s. Women.*

Pilula Rhei c. Hydrargyro.
See Pil. Hydrarg. c Rheo.

Pilula Rhei c. Nuce Vomicâ.
Comp. Rhubarb Pill 3¾ grs. ; Extract of Nux
Vomica ¼ gr. ; Extract of Henbane 1 gr.: in
one pill. *St. Thomas's.*
Comp. Rhubarb Pill 4 grs. ; Extract of Nux Vomica
¼ gr. ; Extract of Henbane 1 gr.: in one pill.
City Chest.

Pilula Rhei Comp. c. Nuce Vomicâ.
Comp. Rhubarb Pill 3½ grs. ; Extract of Nux
Vomica ½ gr.; Capsicum ⅙ gr. ; in one pill.
St. Bartholomew's.

Pilula Scillæ c. Conio.
Extract of Conium 3 grs. ; Compound Squill Pill
2 grs. *Consumption.*
Comp. Pill of Conium 2½ grs. ; Comp. Squill Pill
2½ grs. *Women.*

Pilula Scillæ c. Digitale.
Powdered Squill 2 grs. ; Extract of Conium 2 grs. ;
Powdered Digitalis 1 gr.: in one pill. *Consumption.*

Pilula Scillæ c. Hydrargyro.
See Pilula Hydrargyri c. Scillâ.

Pilula Scillæ c. Ipecacuanhâ.

Comp. Squill Pill $4\frac{1}{2}$ grs.; Ipecacuanha $\frac{1}{2}$ gr.: for one pill. *St. George's.*

Pilula Scillæ c. Morphiâ.

Powdered Squill 2 grs.; Hydrochlorate of Morphia $\frac{1}{8}$ gr.; Ipecacuanha $\frac{1}{2}$ gr.; Oil of Aniseed $\frac{1}{4}$ min.; in one pill. Dose 1 or 2. *Consumption.*

Pilula Scillæ c. Opio.

Squill $\frac{1}{2}$ gr.; Socotrine Aloes 2 grs.; Opium $\frac{1}{2}$ gr.; Canella $\frac{1}{2}$ gr.; Tartar Emetic $\frac{1}{4}$ gr.; Treacle *q. s.*: in one pill. *St. Thomas's.*

Comp. Squill Pill 4 grs.; Opium $\frac{1}{3}$ gr.: in one pill. *Guy's.*

Pilula Sedativa.

Opium $\frac{1}{2}$ gr.; Ipecacuanha $\frac{1}{2}$ gr.; Extract of Henbane 3 grs. *Gt. Northern.*

Pilula Sodæ Arseniatis.

Arseniate of Soda $\frac{1}{16}$ gr.; Sugar of Milk 1 gr.; Extract of Hop 3 grs. *London Ophthalmic.*

Pilula Stramonii.

Extract of Stramonium $\frac{1}{4}$ or $\frac{1}{2}$ gr.; Extract of Gentian 2 grs.; Liquorice Powder *q. s.*: in one pill. *Consumption.*

Extract of Stramonium $\frac{1}{4}$ gr.; Camphor 1 gr.; Squill 2 grs.; Syrup *q. s.*: in one pill. *St. Mary's.*

Extract of Stramonium $\frac{1}{4}$ gr.; Assafœtida 1 gr.; Camphor $\frac{1}{2}$ gr.: Ipecacuanha $\frac{1}{2}$ gr.; Rectified Spirit *q. s.*; Syrup of Red Poppy *q. s.* *Royal Free.*

Extract of Stramonium $\frac{1}{3}$ gr.; Liquorice Root 2 grs.; Treacle *q. s.*; in one pill. *St. Bartholomew's.*

Extract of Stramonium $\frac{1}{2}$ gr.; Extract of Hop 2 grs.: in one pill. *City Chest.*

Pilula Stramonii Co.

Extract of Stramonium $\frac{1}{2}$ gr.; Pill of Ipecacuanha and Squill 4 grs. *Royal Chest.*

Pilula Terebinthinæ.
Chian Turpentine 5 grs.: in one pill. *Westminster Ophthalmic.*

Pilula Terebinthinæ Co.
Chian Turpentine 3 grs.; Rhubarb 2 grs.: in one pill. *St. Thomas's.*

Pilula Terebinthinæ et Zinci.
Chian Turpentine 4 grs.; Sulphate of Zinc 1 gr.: in one pill. *London.*

Pilula Thymol.
Thymol 1 gr.; Gum Tragacanth 1 gr.; Mastich 1 gr.: Glycerine a sufficiency. *Guy's.*

Pilula Trium Sulphatum.
Sulphate of Quinia, Sulphate of Iron, Sulphate of Zinc, of each 1 gr.; Extract of Gentian 2 grs. *Women.*

Pilula Valerianæ et Belladonnæ.
Valerianate of Zinc 2 grs.; Sulphate of Quinia 1 gr.; Extract of Belladonna ¼ gr.; Confection of Roses 1¾ gr. *Charing Cross.*

Pilula Zinci et Belladonnæ.
Oxide of Zinc 2 grs.; Extract of Belladonna ¼ gr.; Extract of Gentian *q. s. Charing Cross.*

Pilula Zinci Comp.
Valerianate of Zinc 2 grs.; Extract of Aloes 1 gr.; Assafœtida Pill 2 grs. *Women.*

Pilula Zinci c. Hyoscyamo.
Sulphate of Zinc 2 grs.; Extract of Henbane 3 grs.: in one pill. *Middlesex.*
Oxide of Zinc 2 grs.; Extract of Henbane 2 grs.: in one pill. *City Chest. Gt. Northern.*

Pilula Zinci Oxidi.
Oxide of Zinc 1, 2, or 3 grs.; Confection of Roses *q. s.*: in each pill. *St. Mary's.*

Oxide of Zinc 2 grs.; Extract of Belladonna ½ gr.;
Glycerine of Tragacanth a sufficiency. *University.*
Oxide of Zinc 2½ grs.; Extract of Gentian 2 grs.:
in one pill. Dose 1 or 2. *Consumption.*
Oxide of Zinc 3 grs.; Extract of Gentian *q. s.*
Charing Cross.
Oxide of Zinc 3 grs.; Extract of Conium 2 grs.
Westminster.
Oxide of Zinc 3 grs.; Extract of Henbane 2 grs.
Royal Chest.
Oxide of Zinc 4 grs.; Mucilage of Tragacanth *q. s.*:
in one pill. *City Chest.*

Pilula Zinci Oxidi et Belladonnæ.

Oxide of Zinc 2½ grs.; Extract of Belladonna ⅛ gr.;
Extract of Gentian *q. s.*: in one pill. *Consumption.*
Oxide of Zinc 2 grs.; Extract of Belladonna ¼ gr.;
Glycerine of Tragacanth *q. s.*: in one pill.
London.

Pilula Zinci Oxidi c. Hyoscyamo.

Oxide of Zinc 3½ grs.; Extract of Henbane 1½ gr.:
in one pill. Dose 1 or 2. *Consumption.*

Pilula Zinci Oxidi c. Morphiâ.

Oxide of Zinc 2½ grs.; Hydrochlorate of Morphia
⅛ gr.; Extract of Hop 2 grs.: in one pill.
Dose 1 to 3. *Consumption.*

Pilula Zinci Sulphatis.

Sulphate of Zinc 2 grs.; Mucilage of Tragacanth
q. s.: in one pill. *City Chest.*
Sulphate of Zinc 1, 2, or 3 grs.; Confection of Roses
q. s.: in one pill. *St. Mary's.*
Sulphate of Zinc 1 or 2 grs.; Extract of Gentian
3 grs.: in one pill. *London.*

Pilula Zinci Valerianatis.

Valerianate of Zinc 1, 2, or 3 grs.; Confection of
Roses, *q. s.*: in each pill. *St. Mary's.*
Valerianate of Zinc 1 gr.; Compound Pill of Assa-
fœtida 2 grs.: in each pill. *Throat.*
Valerianate of Zinc 2 grs.; Confection of Hips 2
grs. *British Skin.*

Pilula Zinci Valerianatis Co.
> Valerianate of Zinc ½ gr. ; Sulphate of Quinia 1 gr. ; Comp. Rhubarb Pill 1 gr. ; Extract of Gentian 2 grs. : in one pill. *London.*

POTASSA CUM CALCE.

Caustic Potash, Lime, equal parts ; Glycerine, *q. s.* *St. Bartholomew's.*
Caustic Potash and recently burnt Lime equal parts ; Rectified Spirit *q. s.* : to make a paste. *Middlesex.*

POTUS.

Potus Acidi Phosphorici.
> Chlorate of Potash 60 grs. ; Diluted Phosphoric Acid 1 drm. ; Decoction of Barley 20 oz. *Women.*

Potus Imperialis.
> Acid Tartrate of Potash 120 grs. ; Potus Limonis 20 oz. *Women.*
> Acid Tartrate of Potash ¼ oz. ; Lemon Juice 4 drms. : Syrup 4 drms. ; Boiling water 20 oz. *London.*

Potus Limonis.
> The Juice and yellow Rind of one Lemon ; White Sugar 120 grs. ; boiling water 20 oz. *Women.*
> Lemon Juice 2 oz. ; Sugar 2 oz. ; water 20 oz. *London.*

Potus Potassæ Chloratis.
> Chlorate of Potash 60 grs. ; boiling water 20 oz. *London. St. Bartholomew's.*

Potus Potassæ Tartratis Acidæ.
> Acid Tartrate of Potash 60 grs. ; half the fresh peel of a Lemon ; boiling water 20 oz. : infuse in a covered vessel, with occasional stirring till cold, and strain. *St. Bartholomew's.*
> Acid Tartrate of Potash 60 grs. ; Lemon Juice ½ oz. ; boiling water 20 oz. *London.*

PULVERES.

Pulvis Æruginis Co.
Verdigris 1 drm.; Powdered Savin 1 drm. *Royal Free.*

Pulvis Alterativus. (*See also* p. 266.)
Grey Powder 1 gr.; Rhubarb 3 grs.; Bicarbonate of Soda 2 grs. *London Ophthalmic.*
Grey Powder 1 gr.; Rhubarb 2 grs.; Carbonate of Magnesia 2 grs.; Comp. Powder of Cinnamon 1 gr. *St. Thomas's.*
Grey Powder 1 gr.; Rhubarb 1 gr.; Bicarbonate of Soda 3 grs. *Gt. Northern.*

Pulvis Aluminis et Acidi Tannici pro inject.
Alum 2 drms.; Tannic Acid 2 drms.: to be dissolved in 20 oz. of water, and mixed with an equal quantity of warm water immediately before use. *London.*

Pulvis Aluminis c. Amylo. (Ear preparation).
Alum in Powder, Starch in Powder, equal parts. *University.*

Pulvis Aluminis Comp.
Alum 180 grs.; Sulphate of Zinc 60 grs. (90 grs. to 20 oz. for a lotion). *Samaritan.*

Pulvis Aluminis Comp. pro inject.
Alum 2 drms.; Sulphate of Zinc 2 drms.: to be dissolved in 20 oz. of water, and mixed with an equal quantity of warm water immediately before use. *London.*

Pulvis Amygdalæ (Almond Meal).
The Sweet Almond, deprived by expression of its oil, and reduced to fine powder. Or the Bitter Almond, deprived by expression of its fixed oil, and by distillation of its Prussic Acid and Essential Oil, and reduced to fine powder. *British Skin.*

Pulvis Anti-Asthmaticus.
Stramonium Powder 4 drms.; Anise Powder 2 drms.; Nitrate of Potash 2 drms.: mix. Let a portion be ignited, and the fumes inhaled. *Consumption.*

Pulvis Bismuthi c. Amylo.
Subnitrate of Bismuth, Starch in powder, of each ½ oz. : mix them thoroughly. *British Skin.*

Pulvis Bismuthi c. Carbone.
Carbonate of Bismuth 10 grs. ; Powdered Wood Charcoal 10 grs. ; Bicarbonate of Soda 5 grs. For a dose. *University.*

Pulvis Bismuthi Co.
Carbonate of Bismuth 5 grs. ; Carbonate of Magnesia 3 grs. ; Powdered Acacia 2 grs. *Consumption.*

Pulvis Bismuthi et Ferri.
Subnitrate of Bismuth 10 grs. ; Saccharated Carbonate of Iron 10 grs. ; Dose 10 to 20 grs. *London.*

Pulvis Bismuthi et Opii.
Carbonate of Bismuth 10 grs. ; Compound Powder of Ipecacuanha 10 grs. *University.*
Subnitrate of Bismuth 5 grs. ; Compound Powder of Chalk and Opium 5 grs. *Westminster.*

Pulvis Bismuthi et Sodæ.
Carbonate of Bismuth 10 grs. ; Bicarbonate of Soda 10 grs. *University.*

Pulvis Calomel. Co.
See Pulvis Hydr. Subchlor. Co.

Pulvis Calomel. et Jalapæ.
See Pulvis Hydr. Subchlor. c. Jalapâ.

Pulvis Calomel. et Opii.
See Pulvis Hydr. Subchlor. et Opii.

Pulvis Calomel. et Rhei.
See Pulvis Hydr. Subchlor. c. Rheo.

Pulvis Calomel. et Scammonii.
See Pulvis Hydr. Subchlor. c. Scammonio.

Pulvis Calumbæ Co.
Calumba, Rhubarb, of each 12 grs.; Cinnamon 6 grs.; Ginger 4 grs.; Gum Acacia, White Bismuth, Bicarbonate of Soda, of each 20 grs. Dose 10 to 30 grs. *St. George's.*

Pulvis Canellæ Aromaticus.
Canella, Ginger, Long Pepper, equal parts. Dose 5 to 10 grs. *St. George's.*

Pulvis Carminativus.
Rhubarb 12 grs.; Prepared Chalk 6 grs.; Oil of Dill 1 min. *St. Thomas's.*

Pulvis Cinchonæ Co.
Yellow Bark in Powder 15 grs.; Bicarbonate of Soda 9 grs.; Aromatic Powder of Canella 3 grs. *St. George's.*

Pulvis Cinchonæ et Rhei.
Comp. Cinchona Powder 15 grs.; Rhubarb 5 grs. *St. George's.*
Cinchona 15 grs.; Rhubarb 10 grs.; Bicarbonate of Soda 10 grs.; Dose 6, 9, or 12 grs. *Westminster Ophthalmic.*

Pulvis Cinchonæ et Sodæ.
Cinchona 5 grs.; Bicarbonate of Soda 5 grs. Dose 10 to 20 grs. *King's. London Ophthalmic.*

Pulvis Cretæ Aromaticus c. Rheo.
Aromatic Powder of Chalk 20 grs.; Rhubarb 10 grs. *St. Bartholomew's.*

Pulvis Elaterii Salinus.
Extract of Elaterium 1 gr.; Acid Tartrate of Potash 25 grs.; Ginger 4 grains. Dose 5 to 30 grs. *Guy's.*

Pulvis Emeticus.
Ipecacuanha 15 grs.; Tartar Emetic 1 gr. *Consumption. St. Thomas's.*
Ipecacuanha 19 grs.; Tartar Emetic 1 gr. *City Chest.*

Ipecacuanha 20 grs. ; Tartar Emetic ½ gr. *St. Bartholomew's. Guy's. Gt. Northern.*
Ipecacuanha 20 grs. Tartar Emetic 1 gr. *St. George's. London.*
Sulphate of Zinc 20 grs. ; warm water *ad lib. Throat.*

Pulvis Ferri et Sodæ Bicarbonatis.
Saccharated Carbonate of Iron 10 grs. ; Bicarbonate of Soda 10 grs. ; Ginger 5 grs. *Royal Chest.*

Pulvis Glycyrrhizæ Comp.
Liquorice Root 15 grs. ; Senna 15 grs. ; Fennel 7½ grs. ; Precipitated Sulphur 7½ grs. ; Sugar 45 grs. *London.*
Powdered Senna 3 drms. ; Powdered Liquorice 3 drms. ; Powdered Fennel 1½ drm. ; Sublimed Sulphur 1½ drm. ; Sugar 9 drms. ; mix. Dose 1 to 2 drms. *Women.*

Pulvis Glycyrrhizæ et Guaiaci.
Comp. Liquorice Powder (Hosp.) 90 grs. ; Guaiacum Resin 7½ grs. ; Magnesia 15 grs. Dose 1 to 2 drms. *London.*

Pulvis Guaiaci Co.
Powdered Guaiacum Resin 15 grs. ; Precipitated Sulphur 15 grs. ; Carbonate of Magnesia 15 grs. ; Gum Arabic 15 grs. ; Bicarbonate of Potash 22½ grs. Dose ½ to 1 drm. *St. George's.*
Guaiacum Resin 15 grs. ; Sublimed Sulphur 15 grs. ; Carbonate of Magnesia 30 grs. Dose 15 to 30 grs. *London.*

Pulvis Guaiaci et Sulphuris.
Resin of Guaiacum 7½ grs. ; Sublimed Sulphur 7½ grs. ; Carbonate of Magnesia 15 grs. : for one dose. *City Chest.*

Pulvis Hydrargyri et Bismuthi.
Grey Powder 2 grs. ; Subnitrate of Bismuth 2 grs. ; Pulv. Rhei et Sodæ 6 grs. *Westminster.*

Pulvis Hydrargyri c. Cretâ et Belladonnâ.
Grey Powder 2 grs. ; Powder of Belladonna Leaves 1 gr. ; Sugar 2 grs. *London Ophthalmic. University.*

Pulvis Hydrargyri c. Cretâ Co.
Grey Powder 3 grs.; Dover's Powder 3 grs. *Gt. Northern. Westminster.*

Pulvis Hydrargyri c. Cretâ et Rheo.
Grey Powder 1 gr.; Comp. Cinnamon Powder 1 gr.; Carbonate of Magnesia 2 grs.; Rhubarb 2 grs. *City Chest.*
Grey Powder 2 grs.; Rhubarb 4 grs. *London Ophthalmic.*

Pulvis Hydrargyri c. Cretâ et Sacchari.
Grey Powder 1½ gr.; Sugar 13½ grs. *University.*
Grey Powder 2 grs.; Sugar 8 grs. *King's.*

Pulvis Hydrargyri et Jalapæ.
Grey Powder 1 part; Compound Jalap Powder 2 parts: mix. Dose 6, 8, or 10 grs. *Westminster Ophthalmic.*

Pulvis Hydrargyri et Opii.
Grey Powder 5 grs.; Dover's Powder 5 grs. *St. George's.*

Pulvis Hydrargyri et Quiniæ.
Grey Powder 3 grs.; Sulphate of Quinia 1 gr. *Westminster Ophthalmic.*

Pulvis Hydrargyri c. Rheo.
Grey Powder 2 grs.; Bicarbonate of Soda 2 grs.; Powdered Rhubarb 4 grs. *Westminster Ophthalmic.*
Grey Powder 2 grs.; Pulvis Rhei et Sodæ 8 grs. *Westminster.*

Pulvis Hydrarg. Subchlor. Comp.
Calomel 4 drms.; Oxide of Zinc 4 drms.; Sulphate of Quinine ½ drm. For outward use. *Women.*

Pulvis Hydrarg. Subchlor. c. Jalapâ.
Calomel 5 grs.; Jalap 10 grs. Dose 15 grs. *St. Mary's.*
Calomel 2 grs.; Jalap 8 grs.; Aromatic Powder of Canella 2 grs. *St. George's.*

Calomel 2 grs. ; Jalap 8 grs. ; Ginger 2 grs. *Guy's. London. London Ophthalmic. St. Thomas's.*
Calomel 3 grs. ; Jalap 9 grs. *City Chest.*
Calomel 1 gr. ; Jalap 8 grs. ; Ginger 1 gr. *Fever.*

Pulvis Hydrarg. Subchlor. c. Jalapâ Comp.
Calomel 5 grs. ; Comp. Powder of Jalap 40 grs. *Royal Free.*

Pulvis Hydrarg. Subchlor. c. Jalapâ Fort.
Calomel 5 grs. ; Powdered Jalap 15 grs. *Royal Free.*

Pulvis Hydrarg. Subchlor. et Opii.
Calomel 3 grs. ; Dover's Powder 7 grs. *St. George's.*

Pulvis Hydrarg. Subchlor. c. Rheo.
Calomel ¼ gr. ; Rhubarb 5 grs. *Royal Chest.*
Calomel 2 grs. ; Rhubarb 8 grs. ; Comp. Cinnamon Powder 2 grs. *City Chest.*
Calomel 2 grs. ; Rhubarb 8 grs. ; Aromatic Powder of Canella 2 grs. *St. George's.*
Calomel 5 grs. ; Rhubarb 20 grs. ; Ginger 5 grs. *London. London Ophthalmic. Westminster Ophthalmic.*

Pulvis Hydrarg. Subchlor. c. Scammonio.
Calomel ½ gr. ; Comp. Powder of Scammony 4 grs. *Royal Chest.*
Calomel 2 grs. ; Comp. Scammon. Powder 10 grs. *City Chest.*
Calomel 2 grs. ; Scammony 4 grs. ; Sugar 2 grs. *Guy's. St. George's.*
Calomel 2 grs. ; Scammony 8 grs. *St. Thomas's.*
Calomel 3 grs. ; Compound Scammony Powder 9 grs. *Samaritan.*
Calomel 5 grs. ; Scammony 20 grs. ; Ginger 5 grs. *London. London Ophthalmic.*

Pulvis Hydrarg. Subchlor. et Sodæ.
Calomel 1 gr. ; Comp. Cinnamon Powder 5 grs. ; Dried Carbonate of Soda 5 grs. *City Chest.*

Pulvis Iodoformi c. Amylo.
Iodoform in fine powder, Starch in powder, of each ½ oz. : rub them well together. *British Skin.*

Pulvis Iodoformi c. Calaminâ. (Ear preparation.)
Iodoform 30 grs.; Calamine 1 drm.; Starch to 1 oz.
University.

Pulvis Iodoformi et Lycopodii.
Iodoform 1 part; Lycopodium 2 parts. *London.*

Pulvis Ipecacuanhæ c. Antimonio.
See Pulvis Emeticus.

Pulvis Ipecacuanhæ c. Antimonio Fortior.
Ipecacuanha 20 grs.; Tartarated Antimony 2 grs.
Fever.

Pulvis Ipecacuanhæ c. Antimonio Mitior.
Ipecacuanha 12 grs.; Tartarated Antimony ½ gr.
Fever.

Pulvis Ipecacuanhæ c. Antimonio Tart.
See Pulvis Emeticus. *Consumption.*

Pulvis Ipecac. Co. c. Bismutho.
Subnitrate of Bismuth 15 grs.; Dover's Powder 5 grs. *Consumption.*

Pulvis Jalapæ c. Calomel.
See Pulvis Hydr. Subchlor. c. Jalapâ.

Pulvis Jalapæ c. Hydr. Subchlorido.
See Pulvis Hydrarg. Subchlor. c. Jalapâ.

Pulvis Jalapæ et Rhei.
Jalap 10 grs.; Rhubarb 5 grs. *London Ophthalmic.*

Pulvis Jalapæ Salinus.
Jalap 15 grs.; Acid Tartrate of Potash 60 grs.; Ginger 5 grs. *London.*

Pulvis Jalapæ c. Scammonio.
Jalap 3 grs.; Scammony 7 grs. *Westminster Ophthalmic.*

Pulvis Kaolin et Creasoti.
Kaolin in powder 1 oz.; Creasote 16 mins. *University.*

Pulvis Lycopodii (Clubmoss Spores).
The yellow powdery spores of Lycopodium Clavatum. *British Skin.*

Pulvis Magnesiæ Compositus.
Sulphate of Magnesia 90 grs.; Sulphate of Soda 30 grs.; Carbonate of Magnesia 15 grs.; Ginger 5 grs. *Consumption.*

Sulphate of Magnesia, Carbonate of Magnesia, Sublimed Sulphur, Acid Tartrate of Potash; of each equal parts. Dose 1 to 4 drms. *Women.*

Pulvis Magnesiæ Sulphatis Compositus.
Sulphate of Magnesia 60 grs.; Light Carbonate of Magnesia, 5 grs.; Ginger 3 grs. *Middlesex.*

Pulvis Magnes. c. Rheo.
See Pulv. Rhei c. Magnesiâ.

Pulvis Potassæ Sodæ et Calumbæ.
Bicarbonate of Potash 10 grs.; Bicarbonate of Soda 10 grs.; Calumba 5 grs.; Ginger 5 grs. *Royal Chest.*

Pulvis Potassæ et Sodæ Comp.
Bicarbonate of Potash 10 grs.; Bicarbonate of Soda 10 grs.; Ginger 3 grs.; Calumba 10 grs.; Sugar 30 grs.; Spirit of Lemon 5 mins. *London.*

Pulvis Quinæ c. Potassæ Bitartrate.
Sulphate of Quinia 1 gr.; Acid Tartrate of Potash 20 grs. *Samaritan.*

Pulvis Rhei. c. Calomel.
See Pulvis Hydr. Subchlor. c. Rheo.

Pulvis Rhei c. Calumbâ.
Rhubarb 5 grs.; Calumba 10 grs.; Bicarbonate of Soda 10 grs.; Compound Powder of Cinnamon 5 grs. Dose 20 to 40 grs. *London.*

Pulvis Rhei c. Hydrargyro. (*See also* pp. 209, 268.)
    Rhubarb 20 grs.; Grey Powder 10 grs.; Aromatic
        Powder of Chalk 5 grs. Dose 6 grs. *Royal Chest.*
    Rhubarb 12 grs.; Grey Powder 3 grs.; Bicarbonate
        of Soda 5 grs. Dose 5 to 20 grs. *Consumption.*
    Rhubarb 4 grs.; Calomel 1 gr.; Ginger 1 gr. *St.
        Thomas's.*

Pulvis Rhei Co. c. Hydrargyro.
    Comp. Powder of Rhubarb 8 grs.; Grey Powder 2
        grs. *King's.*

Pulvis Rhei c. Hydrargyro c. Cretâ.
    See Pulvis Hydrarg. c. Cretâ et Rheo.

Pulvis Rhei et Jalapæ.
    Rhubarb and Jalap equal parts. *City Chest.*

Pulvis Rhei c. Magnesiâ.
    Rhubarb 10 grs.; Carbonate of Magnesia 20 grs.
        London Ophthalmic.
    Rhubarb 10 grs.; Carbonate of Magnesia 30 grs.;
        Aromatic Powder of Canella 5 grs. Dose 20
        to 30 grs. *St. George's.*

Pulvis Rhei c. Potassæ Sulphate.
    Rhubarb 20 grs.; Sulphate of Potash 40 grs. *St.
        Thomas's. City Chest.*
    Rhubarb 2½ grs.; Sulphate of Potash 7 grs.;
        Ginger ½ gr. *Westminster.*

Pulvis Rhei Salinus.
    Rhubarb 10 grs.; Sulphate of Potash 20 grs.
        Guy's.
    Rhubarb 10 grs.; Sulphate of Potash 30 grs. *London.*

Pulvis Rhei c. Sodâ. (*See also* p. 269.)
    Rhubarb 4 grs.; Bicarbonate of Soda 4 grs.; Ginger 2 grs. *Westminster.*
    Rhubarb 5 grs.; Dried Carbonate of Soda 5 grs.;
        Calumba 10 grs. *Guy's.*
    Rhubarb 5 grs.; Dried Carbonate of Soda 5 grs.
        Gt. Northern.

Rhubarb 5 grs. ; Bicarbonate of Soda 5 grs. ; Comp.
Powder of Cinnamon 5 grs. *Samaritan.*
Rhubarb 6 grs.; Ginger 2 grs.; Bicarbonate of
Soda 12 grs. Dose 5 to 20 grs. *Consumption.*
Rhubarb 10 grs.; Bicarbonate of Soda 10 grs.;
Ginger 2½ grs. *Throat.*
Rhubarb 20 grs.; Bicarbonate of Soda 20 grs.
Dose 10 to 20 grs. *London Ophthalmic.*
Rhubarb 20 grs. ; Bicarbonate of Soda 40 grs. *St. Thomas's.*

Pulvis Sabinæ c. Alumine.

Savin in powder, Alum in powder, equal parts. *Royal Free.*

Pulvis Salium.

Sulphate of Soda 20 grs.; Sulphate of Magnesia 30 grs. ; Chloride of Sodium 2 grs. ; Bicarbonate of Soda 5 grs. *Gt. Northern.*
Sulphate of Soda 30 grs. ; Sulphate of Magnesia 30 grs.; Chloride of Sodium 1 gr.; Bicarbonate of Soda 1 gr. *Guy's.*

Pulvis Salis Thermarum Carolinensium (Carlsbad Salt).

Chloride of Sodium 1 drm. ; Bicarbonate of Soda 2 drms.; Sulphate of Soda ½ oz. Dose 1 to 2 drms. in ½ pint of tepid water. *City Chest.*

Pulvis Santonini. (*See* p. 270.)

Pulvis Scammonii Compositus c. Hydrargyro.

Jalap 5 grs. ; Compound Powder of Scammony 5 grs. ; Rhubarb 5 grs. : Grey Powder 3 grs. ; Ginger 2 grs. Dose 10 to 20 grs. *Consumption.*

Pulvis Scammonii c. Hydrargyro.

See Pulvis Hydrarg. Subchlor. c. Scammonio.

Pulvis Scammonii c. Hydrarg. Subchlor.

See Pulvis Hydrarg. Subchlor. c. Scammonio.

Pulvis Sennæ Comp.
Senna 12 grs.; Liquorice 12 grs.; Sublimed Sulphur 6 grs.; Compound Powder of Cinnamon 6 grs.; Sugar 24 grs. Dose 1 drm. *Consumption.*
Same as Pulvis Liquiritiæ Co. German Pharmacopœia. *London Ophthalmic.*

Pulvis Sennæ c. Sulphure.
Senna 2 oz.; Sublimed Sulphur 2 oz.; Acid Tartrate of Potash 4 oz.; White Sugar 2 oz.; Liquorice Root 2 oz. *Charing Cross.*

Pulvis Sodæ Aromaticus.
Dried Carbonate of Soda 20 grs.; Aromatic Powder of Canella 5 grs. *St. George's.*

Pulvis Sodæ et Hydrargyri. (*See also* p. 270.)
Dried Carbonate of Soda 2½ grs.; Calomel ½ gr.; Aromatic Chalk Powder 5 grs. *Guy's.*

Pulvis Sodæ et Hydrargyri c. Cretâ.
Carbonate of Soda 3 grs.; Mercury with Chalk 1 gr. *Guy's.*

Pulvis Sodæ c. Rheo.
See Pulvis Rhei c. Sodâ.

Pulvis Stramonii Compositus.
Powdered Stramonium 1 oz.; Powd. Anise Seeds ½ oz.; Nitrate of Potash ½ oz.; Bruised Tobacco 30 grs. Burn as much as will lie on a shilling and inhale the smoke. *Middlesex.*

Pulvis Sulphuris Co.
Precipitated Sulphur, Tartrate of Soda and Potash, equal parts. *St. George's.*
Sulphur 30 grs.; Bitartrate of Potash 30 grs.; Ginger Powder 5 grs. *Samaritan.*
Sublimed Sulphur ½ oz.; Carbonate of Magnesia 20 grs.; Senna 2 drms.; Acid Tartrate of Potash 1 oz.; Ginger ½ drm. Dose 1 drm. *City Chest.*

Pulvis Talci et Iridis.
Oxide of Zinc 30 grs.; Venetian Talc, in fine powder, sifted, 30 grs.; Florentine Orris, in powder, 30 grs.; Wheat Starch, in powder, 1 oz. Mix them thoroughly. *British Skin.*

Pulvis Tartaratus Co.
Acid Tartrate of Potash 220 grs.; Jalap 20 grs.; Ginger 3 grs. *St. Thomas's.*

Pulvis Terræ Cimoliæ.
(Fuller's Earth, in fine powder.)
Or *Cimolite*, so called from its ancient source, the island of Cimolus, in the Cretan sea. It was also found in the island of Mylos. In this country it is obtained from various districts, viz. Woburn, Redhill, &c. In its natural condition it exists as a solid, compact mass, which consists of an impure silicate of alumina. Its composition is Silica 58; Alumina 25; water 12; Peroxide of Iron 1·30. The iron is now commonly extracted with a view to conferring a better appearance on the powder. *British Skin.*

Pulvis Zinci et Amyli.
Oxide of Zinc 1 oz.; Starch 2 oz.: mix. *Fever. Gt. Northern. University. Westminster.*

Pulvis Zinci Comp.
Oxide of Zinc 1 oz.; Starch or Chalk 1 oz. For outward use. *Women.*

Pulvis Zinci c. Belladonnâ.
Extract of Belladonna ¼ gr.; Oxide of Zinc 5 grs. *Throat.*

Pulvis Zinci et Hydrarg. Subchlor.
Oxide of Zinc 4 parts; Calomel 1 part. For local application. *London.*

Pulvis Zinci Oxidi et Calamini.
Oxide of Zinc ½ oz.; Prepared Calamine ½ oz. *British Skin.*

SANGUIS BOVINUS EXSICCATUS.

Bullocks' Blood deprived of its fibrine and dried at a temperature of about 110° Fahr. *London.*

SAPONES.

Sapo Durus cum Glycerino.

Brand "F. A. Sarg" (of Vienna). *British Skin.*

Sapo Liquidus.

Soft Soap 2 oz.; Rectified Spirit 1 oz.; dissolve. *St. Bartholomew's.*

Sapo Mollis c. Sulphure.

Precipitated Sulphur 120 grs.; Soft Soap 1 oz. Mix thoroughly. *British Skin.*

SOLUTIONES.

Solutio Acidi Citrici.

Citric Acid 30 grs.; water to 1 oz. *London.*

Solutio Acidi Tartarici.

Tartaric Acid 35 grs.; water to 1 oz. *London.*

Solutio Argenti Nitratis.

Nitrate of Silver 60 grs.; Distilled Water 1 oz., No. 1. Nitrate of Silver 20 grs.; Distilled Water 1 oz., No 2. Nitrate of Silver 1 gr.; Distilled Water 1 oz., No 3. *Samaritan.* Nitrate of Silver 1 drm.; Distilled Water 1 oz.: dissolve. *Women.*

Solutio Cantharidis Aceti.

Vinegar of Cantharides 6 drms.; Spirit of Camphor 2 drms. *Consumption.*

Solutio Elaterii Ætherea.

Elaterium 1 gr.; Spirit of Nitrous Ether 1 oz. Dose 1 to 2 drms. *London.*

Solutio Fehling.

Sulphate of Copper $90\frac{1}{2}$ grs.; Distilled Water to 1 oz.: dissolve and keep separate. Tartrate of Potash

364 grs.; Solution of Caustic Soda, sp. gr. 1·12, 4 oz.; water to 5 oz.: dissolve and keep separate. One volume of the Copper Solution, mixed with five volumes of the Soda Solution, and gently warmed, makes the test for grape sugar.

Solutio Ferri Perchloridi et Glycerini.
Perchloride of Iron, Glycerine, equal parts. *Samaritan.*

Solutio Hydrargyri Cyanidi.
Cyanide of Mercury 15 grs.; water to 1 oz. *St. Bartholomew's.*

Solutio Iodoformi.
Iodoform, in very fine powder, 120 grs.; Glycerine to 1 oz. *St. Bartholomew's.*

Solutio Iodi Caustica. vel Pigmentum Iodi.
Iodine 4 drms.; Iodide of Potassium 2 drms.; Rectified Spirit to 2 oz. *Consumption.*

Solutio Terebinthinæ Chiæ Ætherea.
Chian Turpentine 1 oz.; Ether Pure 2 oz. *London.*

SPIRITUS.

Spiritus Ammoniæ Aromaticus.
Carbonate of Ammonia 8 oz.; Strong Solution of Ammonia 4 oz.; Oil of Lemon 2 drms.; Rectified Spirit 5 pints; Distilled Water to 7 pints: dissolve. Same strength as Br. Ph. *St. Bartholomew's.*
Carbonate of Ammonia 6 oz.; Strong Solution of Ammonia 4 oz.; Volatile Oil of Nutmeg 1 drm.; Oil of Lemon 2 drms.; Rectified Spirit 5 pints; Distilled Water to 7 pints: mix. Dose 1 to 2 drms. *London.*

Spiritus Ammoniæ Fœtidus.
Tincture of Assafœtida 6 oz.; Rectified Spirit 3 oz.; Strong Solution of Ammonia 1 oz.: mix. Same strength as Br. Ph. *St. Bartholomew's. London.*

Spiritus Anethi.
Oil of Dill 24 mins.; Rectified Spirit to 1 oz. *London.*

Spiritus Anisi.
Oil of Anise 1 oz.; Rectified Spirit to 50 oz. *St. Bartholomew's.*

Spiritus Carui.
Oil of Caraway 1 oz.; Rectified Spirit to 50 oz. *St. Bartholomew's.*
Oil of Caraway 24 mins.; Rectified Spirit to 1 oz. *London.*

Spiritus Caryophylli.
Oil of Cloves 1 oz.; Rectified Spirit to 50 oz. *t. Bartholomew's.*

Spiritus Cassiæ.
Oil of Cassia 1 oz.; Rectified Spirit to 50 oz. *St. Bartholomew's.*
Oil of Cassia 24 mins.; Rectified Spirit to 1 oz. *London.*

Spiritus Limonis.
Oil of Lemon 24 mins.; Rectified Spirit to 1 oz. *London.*

Spiritus Menthæ. Piperitæ.
Oil of Peppermint 24 mins.; Rectified Spirit to 1 oz. *London.*

Spiritus Menthæ Viridis.
Oil of Spearmint 1 oz.; Rectified Spirit to 50 oz. *St. Bartholomew's.*

Spiritus Pimentœ.
Oil of Pimento 24 mins.; Rectified Spirit to 1 oz. *London.*

Spiritus Saponis Mollis.
Soft Soap 1 oz.; Rectified Spirit ½ fl. oz.: dissolve. *British Skin.*

SUPPOSITORIA.
(*See* also PESSI.)

Basis pro Suppositoriâ.
The same as that used for Pessaries. *London.*

Suppositorium Acidi Carbolici.
Carbolic Acid (No. 2 Calvert) 1 or 2 grs.; Basis to 15 grs.: for one suppository. *London.*

Suppositorium Acidi Tannici et Belladonnæ.
Tannic Acid 3 grs.; Extract of Belladonna 2 grs.; Oil of Theobroma to 15 grs. *Westminster.*

Suppositorium Acidi Tannici et Morphiæ.
Tannic Acid 5 grs.; Hydrochlorate of Morphia $\frac{1}{2}$ gr.; Oil of Theobroma to 15 grs. *Westminster.*
Tannic Acid 5 grs.; Hydrochlorate of Morphia $\frac{1}{4}$ gr.; Oil of Theobroma to 20 grs. *Middlesex.*

Suppositorium Acidi Tannici c. Opio.
Tannic Acid 5 grs.; Powdered Opium 1 gr.; Oil of Theobroma 10 grs. *British Skin.*

Suppositorium Belladonnæ.
Alcoholic Extract of Belladonna leaves $\frac{1}{3}$ gr.; Cacao Butter 15 grs. *University.*
Extract of Belladonna $\frac{1}{2}$, 1, or 2 grs.; Basis to 15 grs.: for one suppository. *London.*
Extract of Belladonna $\frac{1}{2}$ gr.; Oil of Theobroma 20 grs. *Middlesex.*
Extract of Belladonna 2 grs.; Oil of Theobroma to 15 grs. *Westminster.*

Suppositorium Belladonnæ c. Morphiâ.
Extract of Belladonna $\frac{1}{2}$ gr.; Hydrochlorate of Morphia $\frac{1}{4}$ gr.; Oil of Theobroma 20 grs. *Middlesex.*

Suppositorium Gallæ Co.
Galls in powder 3 grs.; Opium 1 gr.; Oil of Theobroma to 20 grs. *Middlesex.*

Suppositorium Gallæ c. Opio.
Galls in powder 5 grs.; Opium in powder 1 gr.; Basis to 15 grs.: for one suppository. *London.*

Suppositorium Iodoformi.
> Iodoform 1½ gr.; Oil of Theobroma 15 grs. *Westminster.*

Suppositorium Nutriens.
> Brand's Concentrated Beef Tea 1 drm.; Tincture of Opium *q. s.*; Glycerine 5 mins.; Beef Tea 15 mins. *Women.*

Suppositorium Opii.
> Opium 1 gr.: Oil of Theobroma to 20 grs. *Middlesex.*
> Opium 1 gr.: Cacao Butter to 15 grs. *University.*
> Comp. Soap Pill 10 grs. *Fever.*
> Comp. Soap Pill 12 grs. *St. Bartholomew's. London.*

SYRUPI.

Syrupus Calcis c. Ferro.
> Hypophosphite of Lime 2 grs.; Syrup of Phosphate of Iron ½ drm.; Syrup to 1 drm. *Consumption.*

Syrupus Calcis Hypophosphitis.
> Hypophosphite of Lime 480 grs.; Glycerine 10 fl. oz.; Spirit of Lemon 20 mins.; Distilled Water a sufficiency. Dissolve the Hypophosphite in 10 fl. oz. of the Water, in a water bath; when cold, filter into the Glycerine, and then add the Spirit and sufficient Distilled Water to make 1 pint. Each fl. drm. contains 3 grs. Hypophosphite of Lime. Dose 1 to 2 drms. *London.*

Syrupus Calcis Lactophosphatis.
> Chloride of Calcium 192 grs.; Phosphate of Soda 768 grs.; Lactic Acid, Concentrated 192 mins.; Glycerine 10 fl. oz.; Distilled Water a sufficiency. Dissolve the Chloride and Phosphate separately, unite the solutions, collect the precipitate, wash with distilled water, and dissolve in the Lactic Acid, then add the Glycerine, and sufficient Distilled Water to make 1 pint. Dose 1 drm. *London.*

Syrupus Calcis Sulphocarbolatis.
> Sulphocarbolate of Lime 800 grs.; Glycerine 10 fl. oz.; Spirit of Lemon 20 mins.; Distilled Water to 20 fl. oz. Dose 1 drm. *London.*

Syrupus Croton Chloral.
Croton Chloral 2 grs.; Glycerine 10 mins.; Syrup of Orange to 1 drm. *Consumption.*

Syrupus Extracti Papaveris.
Extract of Poppies 8 grs.; Simple Syrup 1 oz. Dose 1 drm. *Middlesex.*

Syrupus Ferri Hypophosphitis.
Hypophosphite of Iron, recently prepared 480 grs.; Glycerine 10 fl. oz.; Distilled Water a sufficiency. Dissolve the Hypophosphite in 10 fl. oz. of the Water, in a water bath; when cold, filter into the Glycerine, and then add sufficient Distilled Water to make 1 pint. Each fl. drm. contains 3 grs. of Hypophosphite of Iron. Dose 1 to 2 drms. *London.*

Syrupus Ferri et Manganesii Phosphatis.
Phosphate of Iron, recently prepared, 160 grs.; Phosphate of Manganese, recently prepared, 80 grs.; Phosphoric Acid, sp. gr. 1·50, 10 fl. drms.; Glycerine 10 fl. oz.; Distilled Water a sufficiency. Dissolve the Phosphates in the Acid, previously mixed with 2 fl. oz. of the Water; filter the solution into the Glycerine, and then add sufficient Distilled Water to make 1 pint. Each fl. drm. contains 1 gr. Phosphate of Iron.; ½ gr. Phosphate of Manganese. Dose 1 to 2 drms. *London.*

Syrupus Ferri Phosphatis.
Phosphate of Iron, recently prepared 160 grs.; Phosphoric Acid, sp. gr. 1·5, 10 fl. drms. Glycerine 10 fl. oz.; Distilled Water to 1 pint. Mix the Acid with 2 fl. oz. of the Water, and dissolve the Phosphate in the mixture; filter the solution into the Glycerine, and then add the remainder of the Water. Each fl. drm. contains 1 gr. Phosphate of Iron. Dose 1 drm. *London.*

Syrupus Ferri Phosphatis Compositus.
Phosphate of Iron, recently prepared, 160 grs.; Phosphate of Lime 400 grs.; Phosphate of Potash 30 grs; Phosphate of Soda 25 grs.; Phosphoric Acid, sp. gr. 1·50, 2 fl. oz.; Orange-

Flower Water 1 fl. oz. Liquid Cochineal = 80 grs. Cochineal, 4 fl. drms.; Glycerine 10 fl. oz.; Distilled Water a sufficiency. Mix the Acid with 3 fl. oz. of the Water, and dissolve the Phosphates in the mixture; filter the solution; then add the Glycerine, Cochineal, and Orange-Flower Water; and, lastly, sufficient Distilled Water to make 1 pint. Each fl. drm. contains 1 gr. Phosphate of Iron; $2\frac{1}{2}$ grs. Phosphate of Lime, with Potash and Soda. Dose 1 to 2 drms. *London.*

Syrupus Ferri Phosphatis c. Quiniâ et Strychniâ.

Phosphate of Iron, recently prepared 160 grs.; Phosphate of Quinine 160 grs.; Strychnia 5 grs.: Phosphoric Acid, sp. gr. 1·50, 2 fl. oz.; Glycerine 10 fl. oz.; Distilled Water a sufficiency. Mix the acid with 2 oz. of the Water, and dissolve the Phosphates and the Strychnia in the mixture; filter the solution into the Glycerine, and then add sufficient Distilled Water to make 1 pint. Each fl. drm. contains 1 gr. Phosphate of Iron; 1 gr. Phosphate of Quinine; $\frac{1}{32}$ gr. Strychnia. Dose 1 drm. *London.*

Syrupus Hypophosphitum Compositus.

Hypophosphite of Iron, recently prepared, 160 grs.; Hypophosphite of Manganese, recently prepared, 80 grs.; Hypophosphite of Lime 160 grs.; Hypophosphite of Potash 60 grs.; Hypophosphorous Acid 1 fl. oz.; Glycerine 10 fl. oz.; Distilled Water a sufficiency. Dissolve the Hypophosphites in 8 oz. of the Water, previously mixed with the Acid, in a water bath; filter the solution; allow to cool, and then mix with the Glycerine and sufficient Distilled Water to make 1 pint. Each fl. drm. contains 1 gr. Hypophosphite of Iron; $\frac{1}{2}$ gr. Hypophosphite of Manganese; 1 gr. Hypophosphite of Lime; $\frac{1}{2}$ gr. Hypophosphite of Potash. Dose 1 to 2 drms. *London.*

Syrupus Morphiæ Hydrochloratis.

Hydrochlorate of Morphia $\frac{1}{16}$ gr.; Spirit of Nutmeg 2 mins.; Syrup 30 mins.; water to 1 drm. Dose 1 to 2 drms. *Consumption.*

Syrupus Morphiæ Hydrochloratis Fort.
: Hydrochlorate of Morphia $\frac{1}{8}$ gr.; Syrup of Red Poppies 30 mins.; Spirit of Chloroform 5 mins.; water to 1 drm. *Consumption.*

Syrupus Morphiæ c. Potassæ Chlorate.
: Acetate of Morphia $\frac{1}{12}$ gr.; Chlorate of Potash 2 grs.; Glycerine 10 mins.; Syrup to 1 drm. *Consumption.*

Syrupus Papaveris.
: Made with Treacle instead of Sugar. Same strength as Br. Ph. *St. Bartholomew's.*

Syrupus Rhœados.
: Dried Red-Poppy Petals 13 oz.; Treacle $5\frac{1}{4}$ lbs.; Boiling Distilled Water $5\frac{1}{2}$ pints: infuse the petals in the water for 12 hours, strain, press, and add the treacle. *St. Bartholomew's.*

Syrupus Scillæ c. Morphiâ.
: Oxymel of Squill 30 mins.; Acetate of Morphia $\frac{1}{16}$ gr.; Dil. Hydrocyanic Acid 2 mins.; water to 1 drm. Dose 1 to 2 drms. When prescribed for out-patients, half the dose of Hydrocyanic Acid is given. *Consumption.*

Syrupus Sodæ Hypophosphitis.
: Hypophosphite of Soda 480 grs.; Glycerine 10 fl. oz.; Spirit of Lemon 20 mins.; Distilled Water to 20 fl. oz. Each fl. drm. contains 3 grs. Hypophosphite of Soda. Dose 1 to 2 drms. *London.*

TINCTURÆ.

Tinctura Actææ Racemosæ.
: Actæa Root 5 oz.; Proof Spirit 20 oz.; Macerate 14 days and strain. Dose 20 to 60 mins. *London.*

Tinctura Benzoini Simplex.
: The "Compound Tincture of Benzoin" B. P. without the Aloes. *London.*

Tinctura Chloro-Morphiæ.
Chloroform 6 drms.; Hydrochlorate of Morphia 16 grs.; Dil. Hydrocyanic Acid 3 drms.; Oil of Peppermint 1 drm.; Rectified Spirit 3 oz.; Treacle to 8 oz. Dose 15 to 30 mins. *City Chest.*

Tinctura Cinchonæ.
Pale Cinchona Bark 4 oz.; Proof Spirit 20 oz. Macerate 7 days and strain. Dose ½ to 2 drms. *London.*

Tinctura Cuspariæ.
Cusparia bruised ½ drm.; Proof Spirit 1 oz. *St. George's.*

Tinctura Eucalypti Globuli.
Eucalyptus Leaves 4 oz.; Proof Spirit 20 oz. Macerate 10 days. Dose ½ to 2 drms. *London.*

Tinctura Ferri Sesquichloridi.
Sesquioxide of Iron 1½ oz.; Hydrochloric Acid 5 oz.; Rectified Spirit 15 oz.; Mix the Sesquioxide with the Acid, and digest in a sand-bath, stirring occasionally till the Iron is dissolved: add the Spirit to the solution when cold, and strain. *St. George's.*

Tinctura Gelsemii.
Gelsemium Root 2 oz.; Rectified Spirit 20 oz. Prepared by maceration and percolation. Dose 5 to 20 mins. *London.*

Tinctura Gentianæ.
Gentian sliced ½ drm.; Canella 22½ grs.; Proof Spirit 1 oz. *St. George's.*

Tinctura Hamamelis Virginicæ.
Hamamelis Bark bruised 2 oz.; Proof Spirit 20 oz.: Macerate seven days, and strain. Dose 2 to 5 mins. *London.*

Tinctura Hellebori Nigri.
Black Hellebore Root, in coarse powder 2½ oz.; Proof Spirit 20 oz.: Prepared by maceration and percolation. Dose 20 to 60 mins. *London.*

Tinctura Jaborandi.
Jaborandi Leaves in coarse powder 10 drms.; Rectified Spirit 20 oz.; Prepared by maceration and percolation. Dose 15 to 30 mins. *London.*

Tinctura Podophylli Resinæ.
Resin of Podophyllum 4 grs.; Strong Tincture of Ginger 1 fl. drm.; Rectified Spirit to 1 fl. oz. Dose 15 to 60 mins. *London.*
Resin of Podophyllin 8 grs.; Rectified Spirit of Wine 1 fl. oz. *Guy's*

Tinctura Pruni Virginianæ.
Wild Cherry Bark bruised 2 oz.; Proof Spirit 20 oz. Prepared by maceration and percolation. Dose 15 to 60 mins. *London.*

Tinctura Quillaiæ Saponariæ.
Quillaia Bark (inner), bruised 5 oz.; Alcohol (90 per cent.) 20 oz. Heat to ebullition and filter. Used as an emulsifying agent. *London.*
Quillaia Bark in coarse powder 4 oz.; Rectified Spirit 20 oz. *Guy's.*

Tinctura Sennæ Co.
Senna 60 grs.; Jalap 7½ grs.; Ginger 3¾ grs.; Proof Spirit 1 oz. *St. George's.*

TROCHISCI.

N.B. Lozenges of the Consumption Hospital are made with Acacia Paste.
Lozenges of the London Hospital are made with Black Currant Paste.
Lozenges of the Throat Hospital are made with both Black Currant and Red Currant Paste (with the exception of Carbolic Acid Lozenges and Marshmallow Lozenges). The formulæ for these are omitted, as they are now made and sold by the lozenge manufacturers.

Trochisci Acaciæ.
Consumption.

Trochisci Acidi Benzoici.
About ½ gr. Benzoic Acid in each. One every 4 hours as a stimulant and voice lozenge. *Throat. Gt. Northern.*
Benzoic Acid 1 gr. in each lozenge. *London.*

Trochisci Acidi Carbolici.
About 1 gr. Carbolic Acid in each. One 4 or 5 times a day. Antiseptic and stimulant. *Throat.*

Trochisci Acidi Tannici.
1½ gr. Tannic Acid in each. One every 3 or 4 hours. Strongly astringent. *Throat.*

Trochisci Aconiti.
½ min. Tincture of Aconite in each. One every half-hour or hour. *Throat.*

Trochisci Altheæ.
One every half-hour or hour. Emollient. *Throat.*

Trochisci Ammonii Chloridi.
About 2 grs. Chloride of Ammonium in each. One every 3 hours. *Throat.*

Trochisci Boracis.
3 grs. Borax in each. One every 3 or 4 hours. *Throat.*

Trochisci Catechu.
2 grs. Catechu in each. One every 3 hours. Astringent. *Throat.*

Trochisci Cubebæ.
About ½ gr. Cubebs in each. One every 3 or 4 hours. Useful in bronchitis. *Throat.*
Closely resembles Brown's Bronchial Troches.

Trochisci Glycyrrhizæ.
Extract of Liquorice 3 grs., Oil of Aniseed ⅓ min., in each. *Consumption.*

Trochisci Guaiaci.
4 grs. Guaiacum Resin in each. Dose 1 to 6. *London.*
2 grs. Guaiacum in each. One every 2 hours in acute inflammation, and 3 times a day in chronic affections. *Throat.*

Trochisci Gummi Rubri.
Red Gum 2½ grs., Tincture of Capsicum ⅓ min., in each. Dose 1 to 6. *London.*

Trochisci Kino.
 2 grs. Kino in each. One every 3 or 4 hours.
 Astringent. *Throat.*

Trochisci Krameriæ.
 3 grs. Ext. Rhatany in each. One every 3 or 4 hours. Astringent, and less constipating than Catechu. Dose 1 to 6. *London. Throat.*

Trochisci Lactucæ.
 1 gr. Ext. Lettuce in each. One every hour or two. Soothing. *Throat.*

Trochisci Nitro-Glycerini.
 Contains $\frac{1}{100}$ gr. Nitro-Glycerine. Dose 1 lozenge. *London.*

Trochisci Opii c. Ipecacuanhâ.
 Extract of Opium $\frac{1}{10}$ gr., Ipecacuanha $\frac{1}{3}$ gr., Oil of Aniseed $\frac{1}{2}$ min., in each. *Consumption.*

Trochisci Potassæ Chloratis.
 3 grs. Chlorate of Potash in each. One every 3 or 4 hours. Stimulant and antiseptic: useful in thrush and aphthous ulceration. *Gt. Northern. Throat.*

Trochisci Potassæ Chloratis Acidi.
 Chlorate of Potash 3 grs., Tartaric Acid $1\frac{1}{2}$ gr., in each: flavoured with Essence of Lemon. *Consumption.*

Trochisci Potassæ Citratis.
 3 grs. Citrate of Potash in each. One every 3 or 4 hours. Topical Sialogogue. *Throat.*

Trochisci Potassæ Tartrat. Acidæ.
 3 grs. Acid Tartrate of Potash in each. One every 2 or 3 hours. Topical Sialogogue. *Throat.*

Trochisci Pyrethri.
 1 gr. Pellitory in each. One every 2 or 3 hours. A valuable Sialogogue. *Throat.*

Trochisci Sedativi.
 Contains $\frac{1}{10}$ gr. of Extract of Opium in each. One every 3 or 4 hours. Sedative for irritative coughs. *Throat.*

UNGUENTA.

The London Ophthalmic Hospital gives the following directions:—All the ointments may be made with Paraffin Ointment. In commerce the substances known as Vaseline and Cosmoline (Unguent. Petrolei) may be taken as representing it.
Cerates are included in this group under the head of Ointments.

Unguentum Acidi Boracici.

Boracic Acid 1 drm.; Lard 1 oz.: mix. *British Skin.*
Boracic Acid 1 drm.; Oil of Almonds 4 fl. drms.; White Wax 1 drm.; Lard 1 drm. *Guy's.*
Boracic Acid (in fine powder) 60 grs.; Petroleum Ointment (Hosp.) 1 oz. *London.*
Boracic Acid, levigated, 1 drm.; Paraffin Wax 2 drms.; Lard 4 drms.; Vaseline 2 drms. *Royal Free.*
Boracic Acid 80 grs.; Ointment of Petroleum 1 oz. *St. Bartholomew's.*
Boracic Acid (in fine powder) 1 or 2 parts; White Wax 1 part; Paraffin 2 parts; Almond Oil 4 parts; Prepared Lard 3 parts: dissolve and mix. *University.*

Unguentum Acidi Carbolici.

(*See also* Vasclinum Carbolatum.)
Carbolic Acid (liquefied) ½ drm.; Ointment of Petroleum (Vaseline) 1 oz. *St. Bartholomew's.*
Carbolic Acid (No. 2 Calvert's) 60 grs.; Rectified Spirit 5 mins.; Petroleum Ointment (Hosp.) 1 oz.: melt the last, and when nearly cold add the first two previously mixed. *London.*
Carbolic Acid 1 drm.; Zinc Ointment 1 oz. *Charing Cross.*

Unguentum Acidi Carbolici Compositum.

Carbolic Acid, Sulphur Ointment, Nitrate of Mercury Ointment, equal parts. *Charing Cross.*
Carbolic Acid (No. 2 Calvert's) 1 part; Nitrate of Mercury Ointment 2 parts; Sulphur Ointment 2 parts. *London.*

Unguentum Acidi Carbolici Comp. Fortius.
Carbolic Acid (No. 2 Calvert's) 1 part; Nitrate of Mercury Ointment 1 part; Sulphur Ointment 1 part. *London.*

Unguentum Acidi Chrysophanici.
Chrysophanic Acid 10 grs.; Ointment of Petroleum (Vaseline) 1 oz. *St. Bartholomew's.*
Chrysophanic Acid 5, 10, or 15 grs.; Petroleum Ointment (Hosp.) 1 oz. *London.*
Chrysophanic Acid 10 grs.; Lard 1 oz. *Charing Cross. St. Mary's.*
Chrysophanic Acid 120 grs.; Lard 1 oz.: heat together, constantly stirring, in a water bath for half an hour; when set mix with pestle and mortar. *British Skin.*

Unguentum Acidi Pyrogallici. (Jarisch's Ointment.)
Pyrogallic Acid 60 grs.; Lard 1 oz.: mix. *British Skin.*

Unguentum Acidi Salicylici.
Salicylic Acid 10 grs.; Ointment of Petroleum 1 oz. *St. Bartholomew's.*
Salicylic Acid 30 grs.; Lard 1 oz.: mix. *British Skin.*

Unguentum Acidi Tannici c. Opio.
Tannic Acid 30 grs.; Powdered Opium 30 grs.; Lard 1 oz.: mix. *British Skin.*

Unguentum Æruginis.
Subacetate of Copper 40 grs.; Prepared Lard 1 oz. *Guy's.*

Unguentum Album.
See Ung. Hydrarg. Ammoniati Dil.

Unguentum Alizarini.
Artificial Alizarine 60 grs.; Lard 1 oz.: mix. *British Skin.*

Unguentum Arsenici Causticum.
Arsenious Acid 30 grs.; Lard 1 oz. *St. Mary's.*

Unguentum Atropiæ.
Atropia 4 grs.; Vaseline 1 oz. *Middlesex.*

Unguentum Atropiæ Sulphatis.
Sulphate of Atropia ½ gr.; Vaseline 1 drm. *Guy's.*
Sulphate of Atropia 4 grs.; distilled water 20 mins.; Vaseline 1 oz. *London.*

Unguentum Balsami Peruviani.
Balsam of Peru 1 fl. drm.; Lard 1 oz. *Middlesex.*
Balsam of Peru 1 oz.; Castor Oil 2 oz. *Fever.*

Unguentum Belladonnæ.
Extract of Belladonna 80 grs.; Glycerine 1 drm.; Prepared Lard 1 oz. *Fever.*

Unguentum Belladonnæ Co.
Extract of Belladonna 80 grs.; water *q. s.* to form a paste; Ammoniated Mercury 40 grs.; Benzoated Lard 360 grs. *London Ophthalmic.*
Extract of Belladonna and Ointment of Iodide of Potassium equal parts. *Middlesex.*
Extract of Belladonna 30 grs.; Camphor 10 grs.; Rectified Spirit *q. s.*; Ointment of Petroleum (Vaseline) 1 oz. *St. Bartholomew's.*

Unguentum Belladonnæ c. Hydrargyro.
See Unguentum Hydrargyri c. Belladonnâ.

Unguentum Bismuthi.
White Bismuth 1 drm.; Lard 1 oz. *British Skin. St. George's.*

Unguentum Bismuthi Oleatis.
Oxide of Bismuth 1 oz.; Oleic Acid 8 oz.; White Wax 3 oz.; Vaseline 9 oz. Heat the Oxide with the Oleic Acid until solution takes place; then add the Wax and the Vaseline; melt, and stir till cold. *London.*

Unguentum Bismuthi et Zinci.
Oxide of Bismuth 40 grs.; Oxide of Zinc 20 grs.; Petroleum Ointment (L. H.) 1 oz. Mix. *London.*

Unguentum Bituminis Compositum.
Compound Solution of Bitumen 1 fl. drm.; Petroleum Ointment (L. H.) 1 oz. Mix. *London.*

Unguentum Bituminis c. Hydrargyro.
Syn.—Compound Petroleum Ointment.
Compound Solution of Bitumen 30 mins.; Ammoniated Mercury 10 grs.; Petroleum Ointment (L. H.) 1 oz. Mix. *London.*

Unguentum Calaminæ. (Turner's Ointment.)
Calamine 160 grs.; White Wax 160 grs.; Olive Oil 1 oz. *St. George's.*
Calamine 120 grs.; Yellow Wax 60 grs.; Olive Oil 2 drms.; Lard to 1 oz. *St. Mary's.*
Prepared Calamine 120 grs; Benzoated Lard 1 oz. *British Skin.*
Calamine 60 grs.; Petroleum Ointment (Hosp.) 1 oz. *London.*

Unguentum Camphoræ.
Camphor 30 grs.; Lard 1 oz. *British Skin.*

Unguentum Citrinum.
Citrine Ointment 1 drm.; Lard 7 drms. *Westminster Ophthalmic*, No. 1.
Citrine Ointment 2 drms.; Lard 6 drms. *Westminster Ophthalmic*, No. 2.

Unguentum Creasoti.
Creasote 6 mins.; Nitrate of Mercury Ointment 30 grs. Lard to 1 oz. *Charing Cross.*

Unguentum Creasoti Forte.
Creasote 6 drms.; White Wax 180 grs.: melt with heat. *British Skin.*

Unguentum Cretæ.
Prepared Chalk 1½ oz.; Lard 1 oz. *Middlesex.*
Prepared Chalk 1½ oz.; Petroleum Ointment (Hosp.) 1 oz. *London.*

Unguentum Cretæ Co.
Prepared Chalk 2 drms.; Olive Oil 1 oz.; Vinegar 90 mins.; Lead Plaster 1½ oz. *St. George's.*

Unguentum Daturinæ.
Sulphate of Daturine 4 grs; Vaseline 1 oz. *London.*
Daturine ½ gr.; Vaseline 1 drm. *Guy's.*

Unguentum Daturinæ c. Hydrargyro.
Red Oxide of Mercury, $\frac{1}{10}$ gr.; Daturine $\frac{1}{10}$ gr.: Vaseline 1 drm. *Guy's*.

Unguentum Diachyli. (Hebra's Ointment.)
Lead Plaster ½ oz.; Vaseline ½ oz.: melt with heat. *British Skin. St. Bartholomew's. London.*
Litharge 120 grs.; Olive Oil 1 fl. oz.; Boil with sufficient water, then add Oil of Lavender 5 mins. *Guy's*.
Lead Plaster, Linseed Oil, equal parts. *Charing Cross*.

Unguentum Diachylon (Hebra's Formula).
Olive Oil 15 oz.; Litharge 3¾ oz.: boil to a soft ointment, and add Oil of Lavender 2 drms. *King's*.

Unguentum Eserinæ.
Sulphate of Eserine 2 grs.; Vaseline 1 oz. *London*. Eserine ⅕ gr.; Vaseline 1 drm. *Guy's*.

Unguentum Flavum.
Yellow Oxide of Mercury 8 grs.; Lard 8 drms. *Westminster Ophthalmic*.

Unguentum Flavum c. Atropiâ.
Sulphate of Atropia 4 grs.; Yellow Oxide of Mercury 8 grs.; Vaseline 1 oz. *Gt. Northern*.

Unguentum Hydrarg. Ammoniati Dil.
Ammoniated Mercury 8 grs.; Simple Ointment 1 oz. *London Ophthalmic*.
Ammoniated Mercury 12 grs.; Olive Oil 12 mins.; Lard 1 oz. *University*.
Ammoniated Mercury 15 grs.; Glycerine 1 drm.; Simple Ointment 1 oz. *St. Mary's*.

Unguentum Hydrarg. Ammon. Mitius.
Ammoniated Mercury Ointment 3 drms.; Lard 5 drms. *Charing Cross*.

Unguentum Hydrarg. Ammon. c. Sulphure.
Ammoniated Mercury 6 grs.; Sublimed Sulphur 30 grs.; Benzoated Lard 1 oz. *London Ophthalmic*.

Unguentum Hydrarg. c. Belladonnâ.
Belladonna Ointment, Mercurial Ointment, equal parts. *Middlesex*.
Extract of Belladonna 1 drm. ; Mercurial Ointment 7 drms. *London*.
Extract of Belladonna 40 grs.; Mercurial Ointment 1 oz. *Gt. Northern*.
Extract of Belladonna 2 drms. ; water *q. s.* to form a paste ; Ointment of Mercury 6 drms. *London Ophthalmic*.
Belladonna Ointment, Mercurial Ointment, Ointment of Iodide of Potassium, equal parts. *St. Bartholomew's*.
Extract of Belladonna 10 grs. ; Ammoniated Mercury 5 grs. ; Lard 1 drm. *King's*.

Unguentum Hydrarg. Bibromidi.
Bibromide of Mercury 4 grs. ; Vaseline 1 oz. *Women*.

Unguentum Hydrarg. Camphoratum.
Camphor 60 grs. ; Proof Spirit *q. s.* ; Mercurial Ointment 1 oz. *St. George's*.

Unguentum Hydrarg. Co.
Mercurial Ointment 2 oz.; Soap Cerate Plaster 2 oz. ; Camphor ½ oz. *London*.

Unguentum Hydrarg. Detergens.
Ammoniated Mercury 15 grs.; Liquor Carbonis Detergens 60 mins. ; Lard to 1 oz. *Charing Cross*.

Unguentum Hydrarg. c. Iodo.
Mercurial Ointment, Iodine Ointment, equal parts. *St. George's*.

Unguentum Hydrarg. et Potassii Iodidi. (Lutz's Ointment.)
Red Iodide of Mercury 5 grs. ; Iodide of Potassium 5 grs. ; water *q. s.*; Prepared Lard 1 oz. Triturate the Iodides together in a mortar, add the water drop by drop until a colourless solution is formed, then add the lard and mix thoroughly. *British Skin*.

Unguentum Hydrarg. Iodidi Album.
Corrosive Sublimate 6 grs. ; Iodide of Potassium 20 grs.; Prepared Lard 1 oz. ; water *q. s. St. George's*.

Unguentum Hydrarg. Iodidi Rubri.
Red Iodide of Mercury 1 drm. ; Simple Ointment 9 drms.: mix. This is three times the strength of the Br. Ph., and is used for bronchocele. *Throat.*

Unguentum Hydrarg. Iodidi Rubri Forte.
(Rochard's Ointment.)
Red Iodide of Mercury 60 grs. ; Lard 1 oz. : mix. *British Skin.*

Unguentum Hydrarg. Mitius.
Mercurial Ointment ½ oz. ; Lard 1 oz. *Middlesex. Royal Free.*
Strong Mercurial Ointment 2 drms. ; Camphor 10 grs.; Spirit of Wine *q. s.*; Lard to 1 oz. *King's.*
Mercurial Ointment 1 oz. ; Petroleum Ointment (Hosp.) 2 oz. *London.*

Unguentum Hydrarg. Nitratis Dilutum.
Nitrate of Mercury Ointment 1½ drm. ; Lard 6½ drms. *Westminster.*
Nitrate of Mercury Ointment ⅛, ¼, and ½ oz. ; Lard to 1 oz. *St. Bartholomew's.*

Unguentum Hydrarg. Nitratis Mite.
Ointment of Nitrate of Mercury 1 drm.; Paraffin Ointment (Vaseline) to 1 oz. *London Ophthalmic.*
Ointment of Nitrate of Mercury 1 oz. ; Petroleum Ointment 2 oz. *London.*

Unguentum Hydrarg. Nitrat. Mitius.
Ointment of Nitrate of Mercury 1 drm. ; Paraffin Ointment (Vaseline) to 12 drms. *London Ophthalmic. Gt. Northern.*
Ointment of Nitrate of Mercury 1 drm. ; Lard 8 drms. *Middlesex.*
Ointment of Nitrate of Mercury 1 drm.; Simple Ointment 7 drms. *St. George's.*
Ointment of Nitrate Mercury 34 grs. ; Lard 1 oz. *Guy's.*
Ointment of Nitrate of Mercury 1 oz. ; Vaseline 3 oz. *Royal Free.*

Unguentum Hydrarg. Oleatis.
Liniment of Oleate of Mercury (10 p. c.) 120 grs. ; Lard 1 oz. : mix. *British Skin.*

Unguentum Hydrarg. Oleatis (20 per cent.).
Yellow Oxide of Mercury 2 drms.; Oleic Acid 10 drms. For mode of preparation see Linimentum Hydr. Oleatis. *Throat.*

Unguentum Hydrarg. c. Opio.
Opium 1 drm.; Mercurial Ointment 1 oz. *Gt. Northern.*
Opium 1 drm.; Mercurial Ointment 8 drms. *Middlesex.*
Opium 1 drm.; Mercurial Ointment 11 drms. *Westminster Ophthalmic.*

Unguentum Hydrarg. Oxidi Cinerei.
Black Oxide of Mercury 1 drm.; Lard 1 oz. *St. George's.*

Unguentum Hydrarg. Oxidi Flavi.
Yellow Oxide of Mercury 1 gr. Vaseline 1 drm. *Royal Free.*
Yellow Oxide of Mercury 8 grs.; Vaseline 1 oz. *Middlesex.*
Yellow Oxide of Mercury 8 grs.; Paraffin Ointment (Vaseline) 1 oz. *London Ophthalmic. Gt. Northern.*
Yellow Oxide of Mercury 10 grs.; Vaseline 1 oz. *London.*
Yellow Oxide of Mercury 15 grs.; Benzoated Lard 1 oz.: mix. *British Skin.*
Yellow Oxide of Mercury 2 grs.; Vaseline 1 drm. *Guy's.*

Unguentum Hydrarg. Oxid. Flav. c. Atropiâ.
Sulphate of Atropia 4 grs.; Distilled Water 20 mins.; Yellow Oxide of Mercury Ointment 1 oz. *London.*
Atropia 4 grs.; Yellow Oxide of Mercury 8 grs.; Vaseline 1 oz. *Middlesex.*

Unguentum Hydrarg. Oxidi Rubri.
Eye Ointments.
No. 1. 1 oxide to 7 of Lard.
,, 2. 1 ,, ,, 16 ,,
,, 3. 1 ,, ,, 60 ,, *University.*

Unguentum Hydrarg. Oxidi Rubri Dilut.
(*See also* Unguentum Rubrum.)
Red Oxide of Mercury Ointment 1 drm.: Paraffin Ointment 7 drms. *London Ophthalmic. Westminster.*

Unguentum Hydrarg. Oxidi Rubri Mitius.
Ointment of Red Oxide of Mercury 2 drms.; Lard 6 drms. *King's.*
Ointment of Red Oxide of Mercury $\frac{1}{4}$ oz.; Lard 1 oz. *Middlesex.*
Red Oxide of Mercury 10 grs.; Lard to 1 oz. *Charing Cross.*

Unguentum Hydrarg. Perchloridi.
Perchloride of Mercury 2 grs.; Lard to 1 oz. *Charing Cross.*
Perchloride of Mercury 3 grs.; Lard 1 oz. *St. Mary's.*
Perchloride of Mercury 5 grs.; Petroleum Ointment (Hosp.) 1 oz. *London.*
Perchloride of Mercury 8 grs.; Lard 1 oz.: mix. *British Skin.*

Unguentum Hydrargyri Subchloridi.
Calomel 1 drm.; Lard 1 oz. *St. Mary's.*
Calomel 60 grs.; Petroleum Ointment (Hosp.) 1 oz. *London.*

Unguentum Hydrargyri Sulphatis Flavæ.
Synonyms.—Turbith or Turpeth Mineral Ointment— Bazin's Ointment.
Yellow Sulphate of Mercury 15 grs.; Benzoated Lard 1 oz. Mix thoroughly. The Yellow Sulphate of Mercury (3 HgO,SO$_3$) is thus prepared :—Mercury 1 oz. ; Sulphuric Acid, 1$\frac{1}{2}$ fl. oz. Mix them in' a glass vessel and boil by means of a sand-bath until a dry white mass (HgO,SO$_3$) remains. Reduce this to fine powder, and throw it into boiling water. Pour off the supernatant liquor, and wash the Yellow Precipitate with hot water, and dry it. *British Skin.*

Unguentum Hydrargyri Sulphidi Rubri.
Red Sulphide of Mercury 6 grs. ; Red Oxide of Mercury 6 grs.; Creasote 2 mins.; Lard 1 oz. *St. Bartholomew's.*

Unguentum Hydrarg. Sulphidi Rubri Mitius.
Red Sulphide of Mercury 3 grs.; Red Oxide of Mercury 3 grs.; Creasote 1 min.; Lard 1 oz. *St. Bartholomew's.*

Unguentum Iodoformi.
Iodoform 20 grs.; Petroleum Ointment (Hosp.) 1 oz. *London*.
Iodoform ½ drm.; Simple Ointment 1 oz. *Guy's*.
Iodoform 1 drm.; Vaseline 7 drms. *Royal Free*.
Iodoform 1 drm.; Lard 1 oz.: mix. *British Skin*.
Iodoform 2 drms.; Ointment of Petroleum 1 oz. *St. Bartholomew's*.

Unguentum Iodoformi Comp.
Iodoform 120 grs.; Carbolic Acid (No. 2. Calvert) 6 grs.; Extract of Conium 90 grs.; Petroleum Ointment (Hosp.) 1 oz. *London*.

Unguentum Liq. Carbonis Deterg.
Liq. Carbonis Detergentis 2 fl. drms.; Lard 1 oz. *Guy's*.

Unguentum Metallorum. Vel Ung. Hydrarg. Plumbi et Zinci.
Zinc Ointment, Nitrate of Mercury Ointment, Acetate of Lead Ointment, equal parts. *Guy's. Gt. Northern*.
Zinc Ointment, Nitrate of Mercury Ointment, Sub-acetate of Lead Ointment, equal parts. *Charing Cross*.

Unguentum Morphiæ Acetatis.
Acetate of Morphia 1 gr.; Vaseline 1 oz. *Guy's*.

Unguentum Naptholi (Kaposi's Ointment).
Beta-Napthol 60 grs.; Prepared Lard 1 oz. Melt the Lard on a water bath, add to it the Napthol and stir till the Napthol is dissolved, then remove the mixture and stir constantly until it cools. *British Skin*.

Unguentum Olei Betulæ.
Birch Tar (Oleum Betulæ Albæ) 5 drms.; Yellow Wax 120 grs.: melt. *British Skin*.

Unguentum Olei Staphisagriæ. (*See also* p. 242.)
Oil of Stavesacre 1 drm.; Lard 1 oz.: mix. *British Skin*.

Unguentum Opii.
Opium 20 grs. ; Spermaceti Ointment 1 oz. *St. George's.*
Opium ¼ oz. ; Lard 2 oz. *Middlesex.*
Extract of Opium 30 grs. ; Lard 1 oz. *Guy's.*

Unguentum Petrolei.
Yellow Wax 30 grs. ; Vaseline 1 oz. *London.*

Unguentum Petrolei c. Zinco.
Oxide of Zinc 30 grs.; Ointment of Petroleum 1 oz. *St. Bartholomew's.*

Unguentum Picis Co.
Tar Ointment, Zinc Ointment, and Comp. Subacetate of Lead Ointment, equal parts. *St. Mary's.*
Tar, Vaseline, equal parts. *Gt. Northern.*

Unguentum Picis Juniperi.
Juniper Tar 2 drms. ; Lard 1 oz. *University.*

Unguentum Picis c. Hydrargyro.
Tar Ointment, White Precipitate Ointment, equal parts. *Samaritan.*

Unguentum Picis c. Plumbo.
Tar Ointment and Ointment of Acetate of Lead equal parts. *St. Bartholomew's.*

Unguentum Picis c. Sulphure.
Sublimed Sulphur 2 drms. ; Tar 2 drms; Prepared Chalk 1 drm.; Lard 2 drms. *Middlesex.*

Unguentum Plumbi Acetatis.
Acetate of Lead 12 grs. ; Petroleum Ointment (Hosp.) 1 oz. *London.*

Unguentum Plumbi c. Calaminâ.
Calamine Ointment ½ oz.; Lead Plaster to 1 oz. *St. George's.*

Unguentum Plumbi Carbonatis.
Carbonate of Lead 60 grs. ; Petroleum Ointment 1 oz. *London.*

Unguentum Plumbi Compositum. (Kirkland's Neutral Cerate.)
Lead Plaster 240 grs.; Olive Oil 2 fl. drs.; Prepared Chalk 40 grs.; Dilute Acetic Acid.; 43 mins.; Dissolve the plaster in the oil at a gentle heat. Then add first the chalk, and afterwards the acid, stirring constantly until cold. *British Skin.*

Unguentum Plumbi c. Hydrargyro.
Red Oxide of Lead 45 grs.; Vermilion 25 grs. Prepared Lard 1 oz. *British Skin.*

Unguentum Plumbi c. Hydrargyro Compositum.
Acetate of Lead in fine Powder 10 grs.; Oxide of Zinc 20 grs.; Subchloride of Mercury 10 grs.; Ointment of Nitrate of Mercury 20 grs.; Vaseline to 1 oz. Mix. *London.*

Unguentum Plumbi Glycerini.
Glycerole of Subacetate of Lead ½ drm.; Vaseline 1 oz.: mix. *British Skin.*

Unguentum Plumbi Iodidi.
Iodide of Lead, in fine powder 60 grs.; Petroleum Ointment (Hosp.) 1 oz.; Mix well together. *London.*
Iodide of Lead 1 drm.; Lard to 1 oz. *St. George's.*

Unguentum Plumbi Oleatis.
Oxide of Lead 30 grs.; Oleic Acid ½ oz.; Vaseline to 1 oz.; Dissolve the Oxide in the Oleic Acid with a gentle heat; then add the Vaseline, and stir till cold. *London.*
Lead Plaster, Olive Oil, equal weights. *University.*

Unguentum Plumbi Opiatum.
Solution of Diacetate of Lead 1 drm.; Tincture of Opium 1 drm.; Simple Ointment 1 oz. *St. George's.*

Unguentum Plumbi Oxidi.
Lead Plaster 2 oz.; Olive Oil 3 fl. oz.: mix by a gentle heat. *Middlesex.*

Unguentum Plumbi Oxidi c. Oleo. (*See* Ung. Diachyli.)

Unguentum Plumbi Subacetatis.
Solution of Subacetate of Lead 30 mins.; Petroleum Ointment (Hosp.) 1 oz. Mix. *London.*

Unguentum Plumbi Subacetatis c. Paraffine.
Glycerole of Lead 4½ oz.; Paraffin Ointment, 18 oz.; Paraffin Wax 6 oz.; melt the wax and ointment together, add the glycerole, and stir till cold. *London Ophthalmic.*

Unguentum Rosatum.
Alkanet Root crushed 13 grs.; Otto of Roses 1 min.; White Wax 4 grs.; Prepared Lard 1 oz. Digest the alkanet root for an hour in the lard previously melted on a water-bath, strain the mixture twice through muslin, add the Wax, and after it has melted, stir the mixture while it is cooling until it has assumed a creamy consistency; then add the Otto of Roses, continuing the stirring for a short time. *British Skin.*

Unguentum Rubrum.
Bisulphuret of Mercury 6 grs.; Red Oxide of Mercury 6 grs.; Creasote 2 mins.; Lard 1 oz. *St. Mary's. Samaritan.*
Ointment of Red Oxide of Mercury 1 drm.; Lard 7 drms. *Westminster Ophthalmic.*

Unguentum Rubrum Mitius.
Red Ointment (Hosp.) ½ oz.; Lard ½ oz. *Samaritan.*

Unguentum Sambuci.
Elder leaves 4 drms.; water *q. s.*: Lard 5 drms. Boil till leaves are crisp, and strain with pressure. *St. George's.*

Unguentum Saponis c. Hydrargyro.
Mercurial Ointment 1 oz.; Soap Cerate 1 oz.; Yellow Wax 1 oz.; Camphor 90 grs. *Samaritan.*

Unguentum Simplex.
White Wax ½ oz.; Lard 1 oz. *London Ophthalmic.*
Yellow Wax ¼ oz.; Lard ¾ oz. *St. Bartholomew's.*
Yellow Wax 1 oz.; Lard 1 oz.; Olive Oil 4 fl. oz. *Middlesex.*

Unguentum Staphisagriæ.
(*See also* Ung. Olei Staphisagriæ.)
Stavesacre 1 oz.; Olive Oil 1 oz.; Lard 2 oz. Melt the lard in a water bath, add the oil and stavesacre, maintain the heat for a quarter of an hour, then remove and stir till cold. *London Ophthalmic.*
Stavesacre in powder 1 oz.; Lard 4 oz.: heat for 1 hour, and strain. *Middlesex.* (2 hours) *University.*
Stavesacre seeds 110 grs.; White Wax 55 grs.; Prepared Lard 1 oz. *Guy's.*
Oil of Stavesacre 1 drm.; Lard 1 oz. *St. Mary's.*
Oil of Stavesacre 1 drm.; Ointment of Petroleum 1 oz. *St. Bartholomew's.*

Unguentum Stramonii.
Fresh Stramonium Leaves 1; Lard 4: heat till the leaves become friable, and strain. *Middlesex.*

Unguentum Styracis.
Prepared Storax 2 drms.; Lard 1 oz.: mix. *British Skin.*
Prepared Storax 145 grs.; Rectified Spirit of Wine 3 fl. drms.; Lard 1 oz. *Guy's.*
Prepared Storax liquefied ½ oz.; Lard 1 oz. *St. Bartholomew's.*

Unguentum Sulphuris.
Sublimed Sulphur 1 part.; Petroleum Ointment (Hosp.) 4 parts. Mix. *London.*

Unguentum Sulphuris Co.
Sublimed Sulphur 120 grs.; Carbonate of Potash 60 grs.; Lard 1 oz. *Samaritan.*
Black Sulphur 1 oz.; Green Hellebore Root in powder 1 drm.; Lard 4 oz. *Royal Free.*
Sublimed Sulphur 5 grs.; Ammoniated Mercury 10 grs.; Creasote 3 mins.; Vaseline to 1 oz. *Charing Cross.*
Sublimed Sulphur 30 grs.; Ammoniated Mercury 10 grs.; Red Sulphide of Mercury, 10 grs.; Olive Oil 2 drms.; Creasote 4 mins.; Lard 1 oz. *St. Bartholomew's.*
Sublimed Sulphur ¼ oz.; White Hellebore 30 grs.; Nitrate of Potash ¼ gr.; Soft Soap ¼ oz.; Petroleum Ointment (Hosp.) 1 oz. *London.*

Unguentum Sulphuris c. Hydrargyro.
Sublimed Sulphur 30 grs.; Ammoniated Mercury 5 grs.; Olive Oil 8 mins.; Lard 1 oz. *University*.
Sublimed Sulphur 4 oz.; Ammoniated Mercury ¼ oz.; Æthiops Mineral ¼ oz.; Olive Oil 2 oz.; Lard 8 oz.; Creasote 10 mins. *Samaritan*.

Unguentum Sulphuris Hypochloridi.
Hypochloride of Sulphur 1 drm.; Lard 8 drms. *Middlesex. St. George's*.

Unguentum Sulphuris Iodidi.
Iodide of Sulphur 30 grs.; Prepared Lard 1 oz. *St. George's*.

Unguentum Sulphuris c. Pice.
(Wilkinson's Ointment.)
Precipitated Sulphur 66 grs.; Tar 66 grs.; Prepared Chalk in powder 44 grs.; Soft Soap ½ oz.; Prepared Lard ½ oz.: mix thoroughly. *British Skin*.

Unguentum Sulphuris c. Potassâ.
(Helmerick's Ointment.)
Precipitated Sulphur 120 grs.; Carbonate of Potash 60 grs.; Vermilion 2 grs.; Oil of Bergamot 2 mins.; Lard 1 oz.: mix. *British Skin*.

Unguentum Terebinthinæ Chio.
Chio Turpentine 30 grs.; Vaseline 1 oz. *Women*.

Unguentum Thymol.
Thymol 10 grs. or more; Lard 1 oz. *University*.
Thymol 10 or 20 grs.; Rectified Spirit 10 mins.; Petroleum Ointment (Hosp.) 1 oz. Dissolve the Thymol in the Spirit, and mix thoroughly with the Ointment. *London*.

Unguentum Viride.
Elemi Ointment 1 oz.; Elder Leaf Ointment 1 drm. Copaiba 1½ drm. *St. George's*.

Unguentum Zinci Carbolatum.
Carbolic Acid liquefied 1 drm.; Zinc Ointment 1 oz. *St. Bartholomew's*.

Unguentum Zinci Co.
  Zinc Ointment and Comp. Ointment of Subacetate
    of Lead equal parts. *London Ophthalmic.*
  Carbonate of Lead 1 drm.; Zinc Ointment to 1 oz.
    St. Mary's.
  Zinc Ointment and Chalk Ointment equal parts.
    Middlesex.

Unguentum Zinci c. Hydrargyro.
  Zinc Ointment 6 drms.; Ointment of Nitrate of
    Mercury 1 drm. *St. Thomas's.*

Unguentum Zinci Oleatis.
  Oxide of Zinc 30 grs.; Oleic acid ½ oz.; Vaseline to
    1 oz. Dissolve the oxide in the acid with a
    gentle heat; add the Vaseline and stir till cold.
    London.
  Oleate of Zinc 120 grs.; Lard 1 oz.: mix. *British
    Skin.*
  Oxide of Zinc 1 oz.; Oleic Acid 6 oz.: heat to-
    gether till dissolved, and add Lard 14 oz.: mix.
    University.
  Oxide of Zinc 1 oz.; Oleic Acid 8 oz.: mix and set
    aside for 6 hours; then apply heat to liquefy,
    and form a perfect solution; to this add Paraffin
    Ointment (Vaseline) 9 oz.: dissolve, and stir
    till cold. *London Ophthalmic.*

Unguentum Zinci Oxidi.
  Oxide of Zinc 30 grs.; Petroleum Ointment (Hosp.)
    1 oz. *London.*

Unguentum Zinci c. Plumbo.
  Ointment of Zinc and Ointment of Acetate of
    Lead equal parts. *Guy's.*

VAPORES—INHALATIONES.

(*See also* NEBULÆ *and* FUMI.)

Inhalations prescribed in the Throat Hospital are of five kinds, as follows :—

  Hot: moist air, 130° to 150° F., impregnated with volatile matter.

Cold : moist air 60° to 100° F., impregnated with volatile matter.
Dry : volatile matters vaporized by heat.
Spray : *i.e.* inhalations of atomized fluids.
Fuming : *i.e.* inhalations of the smoke of ignited nitrated papers.

For full particulars respecting these the reader is referred to the Pharmacopœia of the Hospital for Disease of the Throat.

Vapor Acidi Acetici.

Glacial Acetic Acid and Acetic Acid equal parts : mix.
2 teaspoonfuls in a pint of water at 140° F. for each inhalation.
Sedative; used for inflammatory sore throat of scarlet fever. *Throat.*

Vapor Acidi Benzoici.

Benzoic Acid 3 grs. ; Kaolin 12 grs. : rub together and add water ½ oz. ; Tincture of Tolu 18 mins. Shake and make up with water to 1 oz. *Throat.*

Vapor Acidi Carbolici.

Carbolic Acid (No. 2 Calvert) 240 grs. ; Rectified Spirit 4 drms. ; water to 1 oz. ; 1 drm. to 10 oz. water 140° F.
For hot dry inhalation, pour 1 fl. drm. into a heated vessel and let the vapour be inhaled. *London.*
Liquefied Carbolic Acid 30 mins. ; to be added to boiling water 20 oz. *Royal Chest.*
Carbolic Acid 20 grs. ; hot water 1 pint. *City Chest.*
Carbolic Acid Crystals 420 grs. ; water 1 drm. : dissolve.
20 drops in a pint of water at 140° F. for each inhalation.
Antiseptic ; useful in syphilitic ulcerations. *Throat.*

Vapor Acidi Hydrocyanici.

Diluted Hydrocyanic Acid 1 drm. ; water to 1 oz. : mix.
A teaspoonful in a pint of water at 80° F. for each inhalation.
Sedative; useful in cough of laryngeal phthisis. *Throat.*

Vapor Acidi Sulphurosi (Cold).
Sulphurous Acid 1 drm.; water (from 60° to 100° F.) 20 oz. for each inhalation. *Throat.*
Stimulant.

Vapor Ætheris.
Ether, Rectified Spirit, equal parts.: mix.
A teaspoonful in a pint of water at 80° F. for each inhalation.
Sedative and antispasmodic. *Throat.*

Vapor Ætheris Acetici.
Acetic Ether, Rectified Spirit, equal parts.: mix.
A teaspoonful in a pint of Water at 140° F. for each inhalation.
This inhalation may be used as a cold inhalation at 80° F.
Sedative.; often useful in irritation of the larynx. *Throat.*

Vapor Ætheris Chlorici c. Hyoscyamo.
Chloric Ether 30 mins.; Tincture of Henbane 30 mins.; Hot Infusion of Hop, or water, 8 oz. *Consumption.*

Vapor Æthyl Iodidi.
Iodide of Ethyl 5 to 10 drops on a piece of lint or handkerchief for inhalation used in bronchial asthma. *Throat.*

Vapor Aldehydi.
Diluted Aldehyde (15 per cent.) 80 mins.; water to 1 oz. : mix.
A teaspoonful in a pint of water at 140° F. for each inhalation.
Sedative, useful in recent catarrhal congestions; it is contra-indicated in cases of asthma. *Throat.*

Vapor Ammoniæ.
Solution of Ammonia (s. g. ·959) and water, equal parts: mix.
A teaspoonful in a pint of water at 80° F. for each inhalation.
Stimulant; useful in chronic laryngitis and functional aphonia. *Throat.*

Vapor Ammoniæ Benzoatis.

Benzoic Acid 8 grs.; Aromatic Spirit of Ammonia ½ oz.; Spirit of Camphor 3 drms.; Rectified Spirit 1 drm.: a teaspoonful in a pint of water at 80° to 100° F. Use for Valsalvan inhalation. *Throat.*

Vapor Amyl Nitritis.

2 to 5 drops of Nitrite of Amyl on blotting paper or lint for each inhalation. *London.*
Nitrate of Amyl 24 mins.; water to 3 oz. : mix. A teaspoonful in 20 oz. of water at 100° F. for each inhalation.
Antispasmodic; valuable in asthma and spasm of the glottis. *Throat.*

Vapor Benzoini.

Add 1 drm. Simple Tincture of Benzoin (Hosp.) to 10 oz. of water 140° F., and let the vapour be inhaled. *London.*
Comp. Tincture of Benzoin 1 drm. in 20 oz. of water at 140° F. for each inhalation. *Throat. St. Bartholomew's. Gt. Northern.*
Compound Tincture of Benzoin 1 drm.; hot water 1 pint. *City Chest.*
A most valuable sedative inhalation for acute inflammation of the pharynx and larynx.

Vapor Cajuputi.

Oil of Cajuput 8 mins.; Light Carbonate of Magnesia 4 grs.; water to 1 oz. : mix.
A teaspoonful in 20 oz. of water at 140° F. for each inhalation.
Stimulant; useful when the pharyngeal secretion is excessive. *Throat.*

Vapor Calami Aromatici.

Oil of Sweet Flag 5 mins.; Light Carbonate of Magnesia 2½ grs.; water to 1 oz. : mix.
A teaspoonful in 20 oz. of water at 140° F. for each inhalation.
A powerful stimulant in cases of chronic congestion of the larynx. *Throat.*

Vapor Camphoræ.

Spirit of Camphor 10 mins.; Rectified Spirit 20 mins.; water to 1 drm.; to be added to a pint of boiling water. *Royal Chest.*

Spirit of Camphor 2 drms.; hot water 1 pint. *City Chest.*
Spirit of Camphor 1 drm.; Rectified Spirit 3 drms.; water to 1 oz.: mix. *Throat. St. Bartholomew's.*
A teaspoonful in a pint of water at 140° F. for each inhalation. To be inhaled slowly.
Stimulant.

Vapor Cassiæ.

Oil of Cassia 6 mins.; Light Carbonate of Magnesia 3 grs.; water to 1 oz.: mix.
A teaspoonful in 20 oz. of water at 140° F. for each inhalation. *Throat.*
Oil of Cassia 5 mins.; Light Carbonate of Magnesia $2\frac{1}{2}$ grs.; water to 1 oz.; 1 drm to 10 oz. of water at 140° F. *London.*

Vapor Chlori.

Chlorate of Potash, in powder, 1 gr.; Hydrochloric Acid, B.P., 3 mins. Mix in an oz. bottle and let the gas generate and replace the air in the bottle, then put the stopper in the bottle and let it stand for 2 minutes, lastly add gradually, shaking after each addition Distilled Water 1 oz. One oz. in a two-pint jug may be used for an Inhalation (cold). Or, diluted with an equal quantity of water, it may be used as a gargle. Should not be kept prepared above 3 or 4 days. *Throat.*
Stimulant and antiseptic.

Vapor Chloroformi.

Chloroform, Rectified Spirit, of each $\frac{1}{2}$ oz.: mix.
A teaspoonful in 20 oz. of water under 100° F., and another teaspoonful may be added every 5 minutes during the inhalation.
Sedative, used in hay fever and in spasmodic affection of the larynx. *Throat.*
To be used with caution.

Vapor Chloroformi Compositus.

Chloroform 10 mins.; Conium Juice 1 drm.; Glycerine of Carbolic Acid to 2 drms.; boiling water 8 oz. *Consumption.*

Vapor Conii.
Dried Carbonate of Soda 20 grs.; water 20 oz. at 140° F.: dissolve, and add Juice of Conium 2 drms.: for an inhalation. *Throat.*
Carbonate of Soda a teaspoonful; Juice of Conium 2 drms.; Water at 140° F. 10 oz. *London.*
Sedative.

Vapor Conii c. Ammoniâ.
Juice of Conium 2 drms.; Solution of Ammonia 10 mins.; to be added to boiling water 20 oz. *Royal Chest.*

Vapor Creasoti.
Creasote 10 mins.; Light Carbonate of Magnesia 2 grs.: water to 1 drm.; to be added to a pint of boiling water. *Royal Chest.*
Creasote 2 drms.; Light Carbonate of Magnesia 30 grs.; water to 1 oz.: 1 drm. to 10 oz. of water at 140° F. *London.*
Creasote 40 mins.; water 1 oz. *Gt. Northern.*
Creasote ½ oz.; Light Carbonate of Magnesia 90 grs.; water to 3 oz.: mix. *Throat. St. Bartholomew's.*
A teaspoonful in a pint of water at 140° F. for each inhalation.
Stimulant; serviceable for chronic congestion of the larynx.

Vapor Cubebæ.
Oil of Cubebs 2 drms.; Light Carbonate of Magnesia 60 grs.; water to 8 oz.
A teaspoonful in a pint of water at 140° F. for each inhalation.
Valuable stimulant, especially in laryngorrhœa. *Throat.*

Vapor Cubebæ c. Limone.
Oil of Cubebs 1½ drm.; Oil of Lemons ½ drm.; Light Carbonate of Magnesia 1 drm.; water to 3 oz.: mix.
A teaspoonful in 20 oz. of water at 140° F. for each inhalation. *Throat.*

Vapor Hyoscyami.
Henbane seed in powder 60 grs.; percolated with sufficient Proof Spirit to make 1 fl. oz.
A teaspoonful in a pint of water at 140° F. for each inhalation. *Throat.*

Vapor Iodi.
Tincture of Iodine 4 drms.; water to 1 oz.; 1 drm. to 10 oz. water at 140° F.
For hot dry inhalation pour 1 fl. drm. into a heated vessel and let the vapor be inhaled. *London.*
Tincture of Iodine 10 drops for each dry inhalation; may be repeated once or twice during inhalation.
Stimulant, and useful where pus is formed in large quantities. *Throat.*

Vapor Iodi Benzoatus.
Benzoic Acid 16 grs.; Tincture of Iodine 1 oz. Dissolve a teaspoonful in a pint of water at 140° F.
Use for Valsalvan inhalation. *Throat.*

Vapor Iodi Camphoratus.
Tincture of Iodine ½ oz.; strong Solution of Ammonia 30 mins.; Spirit of Camphor 3½ drms.: mix, and after 4 days filter.
A teaspoonful in a pint of water at 80° to 100° F.
Use for Valsalvan inhalation. *Throat.*

Vapor Juniperi Anglici.
English Oil' of Juniper 1 drm.; Light Carbonate of Magnesia 30 grs.; water to 3 oz.: mix.
A teaspoonful to 20 oz. of water at 140° F. for each inhalation.
Stimulant, and very useful in weakness of voice. *Throat.*

Vapor Lupuli.
Hops 1½ oz.; hot water 20 oz. *Consumption.*
Oil of Hop 6 mins.; Light Carbonate of Magnesia 10 grs.; water to 1 oz.: mix.
1 fl. drm. to 10 oz. of water at 140° F. for each inhalation. *London.*

Vapor Lupulinæ.
Lupuline 30 grs., to be added to a pint of water at 140° F. for inhalation. *Throat.*

Vapor Myrti.
Oil of Myrtle 6 mins.; Light Carbonate of Magnesia 6 grs.; water to 1 oz.: mix.
A teaspoonful in 20 oz. water at 140° F. for each inhalation.
Stimulant; very useful in acute tonsillitis. *Throat.*

Vapor Opii.
Extract of Opium 3 grs.; inhale the vapour by means of heat. *Consumption.*

Vapor Origani.
Oil of Common Thyme 5 mins.; Light Carbonate of Magnesia 2½ grs.; water to 1 oz.: mix.
A teaspoonful in 20 oz. of water at 140° F. for each inhalation.
Mild stimulant; useful in subacute inflammations. *Throat.*

Vapor Pini Sylvestris.
Oil of Scotch Pine (Firwool Oil) 40 mins.; Light Carbonate of Magnesia 20 grs.; water to 1 oz.: mix. *Gt. Northern. London. St. Bartholomew's. Throat.*
A teaspoonful in 20 oz. of water at 140° F. for each inhalation. *Throat. St. Bartholomew's.*
1 drm. to 10 oz. water 140° F. for each inhalation. *London.*
Mild stimulant; useful in chronic laryngitis.

Vapor Salviæ.
Oil of Sage 30 mins.; Light Carbonate of Magnesia 15 grs.; water to 3 oz.: mix.
A teaspoonful in 20 oz. of water at 140° F. for each inhalation.
Stimulant. *Throat.*

Vapor Santali.
Oil of Sandal-wood 18 mins.; Rectified Spirit to 3 oz.: mix.
10 or 15 drops to be used with the dry inhaler, and the vapour inhaled: a fresh quantity may be added during the inhalation to make the amount of 1 teaspoonful. It may also be used with Magnesia for steam inhalation.
Sedative; valuable in subacute inflammation with increased mucous secretion. *Throat.*

Vapor Terebenæ.
Terebene pure 40 mins.; Light Carbonate of Magnesia 20 grs.; Distilled Water to 1 oz.
A teaspoonful in a pint of water at 140° F. to be inhaled for ten minutes night and morning.
Stimulant. *Throat.*

Vapor Terebinthinæ.
  Spirit of Turpentine 1 drm.; Tincture of Larch 3
    drms.; hot water to 10 oz. *Consumption.*
  Oil of Turpentine 1 drm.; water 140° F. 10 oz.
    London.

Vapor Thymol.
  Thymol 6 grs.; Rectified Spirit 1 drm.; Light
    Carbonate of Magnesia 5 grs.; water to 1 oz.;
    1 drm. to 10 oz. of water at 140° F. *London.*
  Thymol 18 grs.; Rectified Spirit 3 drms.; Light
    Carbonate of Magnesia 9 grs.; water to 3 oz.:
    mix.
  A teaspoonful in 20 oz. of water at 140° F. for each
    inhalation. *St. Bartholomew's. Throat.*
  A strong stimulant and disinfectant; used in pharyngitis and laryngitis when associated with
    exanthemata.

VASELINA.

Vaselinum Carbolatum.
  Carbolic Acid 20 grs.; Oil of Cloves 16 drops;
    Vaseline 2 oz. *Women.*

Vaselinum c. Eucalypto.
  Oil of Eucalyptus 12 mins.; Olive Oil 3 oz.;
    Vaseline 1 oz. *Women.*

VINA.

Vinum Ferri. (*See also* Liq. Ferri et Ammon. Cit.)
  Tartarated Iron 20 grs.; Sugar ½ drm.; Rectified
    Spirit 1½ drm.; hot water to 2 oz. *City Chest.*
  Iron Filings 60 grs.; Acid Tartrate of Potash ¾ oz.;
    water 32 oz. or *q. s.*; Proof Spirit 20 oz. Rub
    the Iron and the Acid Tartrate together, and
    expose to the air in an open glass vessel for
    6 weeks, with 1 oz. of water, stirring daily
    with a spatula, water being added now and
    then that they may be always moist. Then dry
    with a gentle heat, rub into powder, and mix
    with 30 oz. of water. To the strained liquor
    add the Spirit. *St. George's.*
  Iron Wire *q. s.*; Acid Tartrate of Potash 2 oz.;
    Tartaric Acid 1½ oz.; Dil. Sulphuric Acid 1 oz.;

Glycerine 8 oz. ; Caramel 1 oz. ; Rectified Spirit 80 oz. ; water to make 2 gallons. Mix the ingredients with the water and Spirit, and macerate for 28 days, shaking occasionally, then filter. *University.*

Tartarated Iron 320 grs. ; Sugar 1 oz. ; Rectified Spirit 10 drms. ; water to 20 oz.: mix. *London.*

Tartarated Iron ¼ oz. ; Rectified Spirit 1 oz. ; water to 10 oz. *Gt. Northern.*

Vinum Ipecacuanhæ.

Ipecacuanha 1 oz. ; Sherry Wine 4 oz. ; Rectified Spirit 4 oz. ; water 12 oz. Macerate 7 days, press, and filter, then add water to make 20 oz. *London. St. Bartholomew's.*

Vinum Opii.

Extract of Opium 1 oz. ; Sherry 4 oz. ; Rectified Spirit 4 oz. ; water 12 oz. Mix and filter. Dose 10 to 40 mins. *London.*

Vinum Rhei.

Rhubarb Root 1½ oz. ; Sherry 4 oz. ; Rectified Spirit 4 oz.; water 12 oz. Macerate 7 days. Dose 1 to 2 drms. *London.*

PHARMACOPŒIA

OF THE

HOSPITAL FOR SICK CHILDREN,*

Interspersed with other Formulæ for Children, taken from the General Hospitals, to which the names of each Hospital are attached.

ALL THOSE WITHOUT NAMES BELONG TO THE CHILDREN'S HOSPITAL.

BALNEA.

Balneum Alkalinum.

Carbonate of Soda 1 oz.; hot water at 96° F. 5 gallons.

Balneum Sulphuratum.

Sulphurated Potash 1 oz.; hot water 7 gallons.

DUSTING POWDER.

(*See also* Pulv. Zinci et Amyli.)

Oxide of Zinc 1; Starch Powder 3 : mix. *Children's. St. Mary's. Samaritan.*

GUTTÆ.

Gutta Atropiæ Sulphatis.

Sulphate of Atropia 1 gr.; Distilled Water 1 oz.

---

* The doses of remedies used internally are calculated of a strength suitable to children of one year old for the *Children's Hospital, London*, and *St. Mary's*; one to two years for the *Samaritan*. For the other Hospitals the dose is for two to ten years.

Gutta Podophylli Resinæ.
> Resin of Podophyllin 1 gr.; Rectified Spirit 1 drm. Dose 10 drops.

LINCTUS.

Linctus Infantilis.
> Compound Tincture of Camphor 2½ mins.; Ipecacuanha Wine 2½ mins.; Glycerine 20 mins.; Peppermint Water to 1 drm. *London.*

Linctus Pectoralis.
> Oxymel of Squill 5 mins.; Compound Tincture of Camphor 2½ mins.; Spirit of Nitrous Ether 2½ mins.; water to 1 drm.: mix.

LOTIONES.

Lotio Aluminis.
> Alum 4 grs.; water 1 oz.

Lotio Ammonii Chloridi.
> Chloride of Ammonium 15 grs.; Methylated Spirit 1 drm.; water to 1 oz.

Lotio Belladonnæ.
> Extract of Belladonna 3 grs.; water to 1 oz.

Lotio Boracis Composita.
> Borax 8 grs.; Sulphate of Zinc 8 grs.; Tincture of Opium 8 mins.; water to 1 oz.

Lotio Calaminæ.
> Oxide of Zinc 30 grs.; Sifted Calamine 30 grs.; Glycerine 24 mins.; Lime Water 24 mins.; water to 1 oz.

Lotio Picis Composita.
> Soft Soap, Tar, Methylated Spirit, equal parts.

Lotio Potassæ Chloratis c. Borace.
> Chlorate of Potash 5 grs.; Borax 6 grs.; water 1 oz.

Lotio Rubra.
Sulphate of Zinc 2 grs.; Comp. Tincture of Lavender 10 mins.; water to 1 oz.

Lotio Sodæ Hyposulphitis.
Hyposulphite of Soda 25 grs.; water 1 oz.

Lotio Zinci Chloridi.
Chloride of Zinc 1 gr.; water 1 oz.

Lotio Zinci Oxidi.
Oxide of Zinc 20 grs.; Glycerine 20 mins.; water 1 oz.

Lotio Zinci Sulphatis.
Sulphate of Zinc 2 grs.; water 1 oz.

MISTURÆ.

Mistura Acidi Hydrochlorici.
Diluted Hydrochloric Acid 1½ min.; Syrup 5 mins.; water to 1 drm.
Diluted Hydrochloric Acid ½ drm.; Spirit of Chloroform ½ drm.; Syrup ½ drm.; water to 1½ oz. Dose 1 to 2 drms. *Middlesex.*

Mistura Acidi Nitro-hydrochlorici.
Diluted Nitro-hydrochloric Acid 1½ min.; Spirit of Chloroform 1½ min.; Tincture of Orange 2½ mins.; water to 1 drm.
Diluted Nitro-hydrochloric Acid 1½ min.; Spirit of Chloroform 1½ min.; Syrup of Orange 5 mins.; water 1 drm. *St. Mary's.*

Mistura Ætheris c. Ammoniâ.
Spirit of Ether 3½ mins.; Aromatic Spirit of Ammonia 3½ mins.; Tincture of Orange 2 mins.; Camphor Water to 1 drm.
Carbonate of Ammonia 3 grs.; Syrup of Squill 1½ drm.; Compound Ether Draught to 1½ oz. Dose 1 to 2 drms. *Middlesex.*

Mistura Aluminis.
Alum 2½ grs.; Syrup 20 mins.; water to 1 drm.
Alum 2½ grs.; Syrup 10 mins.; water 1 drm. *St. Mary's.*

Alum 40 grs.; Syrup of Red Poppies 1 drm.;
water to 1½ oz. Dose 1 drm. *Middlesex*.

Mistura Aluminis Composita.
Alum 2½ grs.; Compound Tincture of Camphor 2½
mins.; Dil. Sulphuric Acid 1 min.; Camphor
Water to 1 drm. *London*.

Mistura Ammoniæ Acetatis.
Solution of Acetate of Ammonia 3 drms.; Citrate
of Potash 60 grs.; Glycerine 1 fl. drm.; water
to 1½ oz. Dose 1 to 2 drms. *Middlesex*.

Mistura Ammoniæ Citratis.
Carbonate of Ammonia 1½ gr.; Citric Acid 1½ gr.;
water 1 drm.

Mistura Aperiens.
Extract of Jalap ¼ gr.; Syrup of Senna 10 mins.;
Spirit of Chloroform 1½ mins.; Tincture of
Ginger 1¼ mins.; Syrup 10 mins.; Caraway
Water to 1 drm.

Mistura Aperiens Infantilis.
Sulphate of Magnesia 2½ grs.; Tincture of Rhubarb
10 mins.; Caraway Water to 1 drm. *London*.

Mistura Aromatica.
Aromatic Chalk Powder 5 grs.; Mucilage of Acacia
30 mins.; water 30 mins. *Children's*. (For
Middlesex see the following.)

Mistura Aromatica c. Rheo.
Carbonate of Magnesia 1 gr.; Aromatic Chalk Powder 5 grs.; Tinct. Rhubarb 5 mins.; Peppermint Water to 1 drm. *Children's*. Syn. Mist.
Aromat. *Middlesex*.

Mistura Astringens Infantilis.
Liquid Extract of Bael 20 mins.; Syrup of Red
Gum 10 mins.; Tincture of Opium ½ min.;
Decoction of Logwood to 1 drm. *Samaritan*.

Mistura Belladonnæ.
Extract of Belladonna ⅛ gr.; Spirit of Chloroform 2
mins.; water to 1 drm.

Extract of Belladonna ⅛ gr.; Glycerine 10 mins.;
water to 1 drm. *London*.

Mistura Belladonnæ c. Potassii Bromido.

Bromide of Potassium 4 grs.; Tincture of Belladonna 4 mins.; Glycerine 10 mins.; water to 1 drm.

Mistura Bismuthi Astringens.

Subnitrate of Bismuth 2 grs.; Tannic Acid 1 gr.; Aromatic Powder of Chalk and Opium 1 gr.; Acacia Mixture to 1 drm. *London*.

Mistura Bismuthi Composita.

Subnitrate of Bismuth 2 grs.; Bicarbonate of Soda 2 grs.; Comp. Tragacanth Powder 1½ grs.; Spirit of Chloroform 1 min.; Caraway Water 1 drm.

Mistura Cannabis Indicæ.

Tincture of Indian Hemp. 2¼ mins.; Spirit of Chloroform 1 min.; Mucilage of Acacia 15 mins.; water to 1 drm.

Mistura Carminativa.

Bicarbonate of Soda 1½ gr.; Aromatic Spirit of Ammonia 1½ min.; Glycerine 5 mins.; Peppermint Water to 1 drm.

Bicarbonate of Soda 1½ gr.; Aromatic Spirit of Ammonia 2 mins.; Glycerine 10 mins.; Dill Water to 1 drm. *London*.

Bicarbonate of Soda 1½ grs.; Aromatic Spirit of Ammonia 2 mins.; Aromatic Powder of Chalk 2 grs.; Syrup 10 mins.; water to 1 drm. *St. Mary's*.

Bicarbonate of Soda 2 grs.; Aromatic Spirit of Ammonia 2 mins.; Glycerine 5 mins.; Peppermint Water to 1 drm. *Middlesex*.

Carbonate of Magnesia 20 grs.; Comp. Tincture of Cardamoms 1 drm.; Tincture of Rhubarb 1 drm.; Spirit of Chloroform 6 mins.; Syrup of Ginger 2 drms.; Caraway Water to 1½ oz. Dose 1 drm. *Samaritan*.

Mistura Catechu Aromatica.

Aromatic Chalk Powder 20 grs.; Infusion of Catechu 1½ oz. Dose 1 to 2 drms. *King's*.

Mistura Chloral.
Hydrate of Chloral 30 grs.; Bromide of Potassium 30 grs.; Syrup of Tolu 1½ drm.; Cassia Water to 1½ oz. Dose 1 to 2 drms. *Middlesex.*

Mistura Chloral Hydratis c. Belladonnâ.
Chloral Hydrate 1 gr.; Tincture of Belladonna 1 min.; Glycerine 20 mins.; water to 1 drm. *Charing Cross.*

Mistura Chloral c. Ipecacuanhâ.
Hydrate of Chloral 12 grs.; Ipecacuanha Wine 1 drm.; Syrup of Tolu 2 drms.; water to 1½ oz. Dose 1 drm. *Samaritan.*

Mistura Chloral Sedativa.
Chloral Hydrate 1 gr.; Tincture of Belladonna 2 mins.; Syrup 20 mins.; water to 1 drm. *Samaritan.*

Mistura Chlori.
Chlorate of Potash 20 grs.; Hydrochloric Acid 60 mins.; Syrup 2 oz.; water 18 oz.
Put the Chlorate of Potash and Acid into a bottle, and lightly cork; add water by degrees, shaking the bottle as the water is added. Dose 1 drm.

Mistura Cinchonæ.
Compound Tincture of Cinchona 10 mins.; Syrup 10 mins.; water to 1 drm. : mix.
Tincture of Yellow Bark 2 drms.; Dil. Nitric Acid 30 mins.; Glycerine 1 drm.; water to 1½ oz. Dose 1 to 2 drms. *Middlesex.*

Mistura Cinchonæ Comp.
Comp. Tincture of Cinchona 5 mins.; Dil. Nitric Acid 2 mins.; Syrup 10 mins.; water to 1 drm. *Samaritan.*

Mistura Cretæ c. Catechu.
Tincture of Catechu 10 mins.; Spirit of Chloroform 5 mins.; Chalk mixture to 1 drm.
Tincture of Catechu 5 mins.; Powder of Chalk and Opium 2 grs.; Syrup 10 mins.; water to 1 drm. *Samaritan.*

Mistura Cretæ c. Hæmatoxylo.
Chalk Mixture ½ drm.; Tincture of Catechu 3¾ mins.; Decoction of Logwood to 1 drm. *King's.*

Mistura pro Diarrhœâ Infantum.
Aromatic Powder of Chalk 15 grs.; Compound Tincture of Camphor 8 mins.; Tincture of Catechu 20 mins.; Chalk Mixture to 1 oz. *Royal Free.*

Mistura Extracti Filicis.
Liquid Extract of Male Fern 30 mins.; Spirit of Cinnamon 20 mins.; Syrup 2 drms.; Mucilage of Acacia to ½ oz.
N.B.—This mixture is for children of 2 years old and upwards.

Mistura Ferri.
Solution of Perchloride of Iron 4 mins.; Glycerine 10 mins.; Cinnamon Water to 1 drm. *Samaritan.*

Mistura Ferri Aperiens.
Sulphate of Magnesia 2½ grs.; Sulphate of Iron ½ gr.; Dil. Sulphuric Acid 1 min.; Syrup of Ginger 5 mins.; Dill Water ½ drm.; water to 1 drm. *Charing Cross.*

Mistura Ferri Laxans.
Sulphate of Magnesia 2½ grs.; Dil. Sulphuric Acid 1 min.; Sulphate of Iron ¼ gr.; Syrup of Ginger 2 mins.; Peppermint Water to 1 drm.

Mistura Ferri Perchloridi.
Tincture of Perchloride of Iron 2½ mins.; Syrup 10 mins.; water to 1 drm.

Mistura Ferri et Potassii Bromidi.
Syrup of Phosphate of Iron ½ drm.; Bromide of Potassium 5 grs.; Distilled Water to 2 drms. *Samaritan.*

Mistura Ferri et Potassæ Chloratis.
Solution of Perchloride of Iron 2 mins.; Chlorate of Potash 3 grs.; water to 1 drm. *Charing Cross.*

Mistura Gentianæ Alkalina.

Bicarbonate of Soda 2 grs.; Comp. Infusion of Gentian 1 drm.

Mistura Gummi Rubri.

Liquid Extract of Red Gum 5 mins.; Glycerine 10 mins.; Spirit of Chloroform 5 mins.; Peppermint Water to 1 drm.

Mistura Hæmatoxyli.

Diluted Nitric Acid 1 min.; Compound Tincture of Camphor 2½ mins.; Decoction of Logwood to 1 drm.

Diluted Nitric Acid 1 min.; Tincture of Catechu 5 mins.; Syrup 8 mins.; Decoction of Logwood to 1 drm. *St. Mary's.*

Diluted Nitric Acid 1 min.; Tincture of Catechu 5 mins.; Glycerine 5 mins.; Decoction of Logwood to 1 drm. *London.*

Diluted Nitric Acid 2 mins.; Compound Tincture of Camphor 5 mins.; Syrup of Ginger 10 mins.; Decoction of Logwood to 1 drm. *Samaritan.*

Extract of Logwood 5 grs.; Tincture of Catechu 10 mins.; Caraway Water 25 mins.; hot water 25 mins. Dose 1 to 2 drms. *Middlesex.*

Mistura Hæmatoxyli Composita.

Extract of Logwood 4 grs.; Ipecacuanha Wine 3 mins.; Wine of Opium ½ min.; Chalk Mixture to 2 drms.

Mistura Ipecacuanhæ.

Ipecacuanha Wine 2½ mins.; Glycerine 10 mins.; water to 1 drm.

Ipecacuanha Wine 2½ mins.; Tincture of Squill 2½ mins.; Glycerine 5 mins.; water to 1 drm. *Middlesex.*

Ipecacuanha Wine 5 mins.; Mucilage 15 mins.; Syrup 8 mins.; water 1 drm. *St. Mary's.*

Mistura Ipecacuanhæ Ammoniata.

Ipecacuanha Wine 2½ mins.; Tincture of Squill 2½ mins.; Aromatic Spirit of Ammonia 2 mins.; Glycerine 5 mins.; water to 1 drm. Dose 1 to 2 drms. *Middlesex.*

Mistura Ipecacuanhæ c. Ammoniâ.
Carbonate of Ammonia ½ gr.; Mixture of Ipecacuanha 1 drm.

Mistura Ipecacuanhæ Co.
Ipecacuanha Wine 3 mins.; Comp. Tincture of Camphor 5 mins.; Syrup of Tolu 10 mins.; Spirit of Nitrous Ether 3 mins.; water to 2 drms. *Samaritan.*

Ipecacuanha Wine 20 mins.; Carbonate of Soda 8 grs.; Nitrate of Potash 4 grs.; Spirit of Nitrous Ether 20 mins.; Solution of Acetate of Ammonia ½ oz.; water to 1 oz. *Royal Free.*

Mistura Ipecacuanhæ Opiata.
Ipecacuanha Wine 2½ mins.; Bicarbonate of Soda 2 grs.; Spirit of Nitrous Ether 2½ mins.; Compound Tincture of Camphor 2½ mins.; water to 1 drm.

Bicarbonate of Soda 20 grs.; Sweet Spirit of Nitre 30 mins.; Tincture of Opium 3 mins.; Ipecacuanha Mixture to 1½ oz. Dose 1 drm. *Middlesex.*

Mistura Ipecacuanhæ c. Sodâ.
Ipecacuanha Wine 3 mins.; Bicarbonate of Soda 3 grs.; Dill Water ½ drm.; water to 1 drm. *Charing Cross.*

Ipecacuanha Wine 5 mins.; Bicarbonate of Soda 2 grs.; Aromatic Spirit of Ammonia 2 mins.; Peppermint Water to 1 drm. *Consumption.*

Mistura Liq. Potassæ c. Ipecacuanhâ.
Solution of Potash 30 mins.; Ipecacuanha Wine 30 mins.; Syrup of Red Poppies 2 drms.; Mucilage ½ oz.; water to 2 oz. Dose 1 to 2 drms. *King's.*

Mistura Lobeliæ et Belladonnæ.
Ethereal Tincture of Lobelia 1 min.; Tincture of Belladonna 2 mins.; Almond Mixture to 1 drm. *Charing Cross.*

Mistura Magnesiæ Sulphatis.
Sulphate of Magnesia 2½ grs.; Diluted Sulphuric Acid 1 min.; water 1 drm.

Sulphate of Magnesia 5 grs.; Tincture of Rhubarb 10 mins.; Syrup of Ginger 5 mins.; Caraway Water 1 drm. *Samaritan.*

Mistura Magnes Sulph. c. Ipecacuanhâ.

Sulphate of Magnesia 20 grs.; Nitrate of Potash 10 grs.; Solution of Acetate of Ammonia 2 drms.; Ipecacuanha Wine 1 drm.; Syrup of Lemons ½ oz.; Water to 4 oz. Dose 2 drms. *Samaritan.*

Mistura Magnes. Sulph. c. Rheo.

See Mist. Rhei c. Magnes. Sulph.

Mistura Olei Morrhuæ.

Cod Liver Oil 30 mins.; Tincture of Calumba 5 mins.; Lime Water 30 mins.
Cod Liver Oil 30 mins.; Glycerine 10 mins.; Solution of Lime to 1 drm. *London.*

Mistura Olei Morrhuæ c. Ferro.

Cod Liver Oil ½ drm.; Carbonate of Potash ⅛ gr.; Steel Wine ½ drm.
Cod Liver Oil 20 mins.; Glycerine 20 mins.; Steel Wine 20 mins. *St. Mary's.*
Cod Liver Oil 15 mins.; Gum Acacia 2½ grs.; Tartarated Iron 1½ gr.; Treacle 30 mins.; Cassia Water to 1 drm. *London.*
Cod Liver Oil ½ oz.; Iron Wine 1 oz.; Oil of Cassia 2 mins.; Mucilage 2 drms. Dose 1 drm. *Middlesex.*

Mistura Olei Morrhuæ c. Ferro et Glycerino.

Cod Liver Oil 20 mins.; Steel Wine 20 mins.; Glycerine 20 mins.

Mistura Olei Morrhuæ c. Glycerino.

Cod Liver Oil 20 mins.; Lime Water 20 mins.; Glycerine 20 mins.

Mistura Olei Ricini. (*See also* p. 140.)

Castor Oil 5 mins.; Mucilage of Acacia 15 mins. Syrup 30 mins.; Peppermint Water to 1 drm.
Castor Oil 5 mins.; Mucilage of Acacia 15 mins. Syrup 20 mins; water to 1 drm. *Charing Cross.*

Castor Oil 5 mins.; Mucilage of Acacia 15 mins.;
Syrup 30 mins.; water to 1 drm. *St. Mary's.*
Castor Oil 5 mins.; Tincture of Rhubarb 5 mins.;
Tincture of Quillaia 2 mins.; Glycerine 10
mins.; Pimento Water to 1 drm. *London.*
Castor Oil 7½ mins.; Tincture of Opium ¼ min.;
Powder of Acacia 3¾ grs.; Sugar 3¾ grs.;
water to 1 drm. *King's.*
Castor Oil 7½ mins.; Powdered Gum Acacia 2½
grs.; Sugar 2½ grs.; Caraway Water to 1 drm.
Middlesex.
Castor Oil 1 drm.; Bicarbonate of Potash 10 grs.;
Mucilage 2 drms.; Syrup 1 drm.; Aniseed
Water to 1 oz. *Royal Free.*
Castor Oil 10 mins.; Mucilage of Acacia 15 mins.;
Syrup 20 mins.; Dill Water to 1 drm.
Samaritan.

Mistura pro Pertussi.

Diluted Hydrocyanic Acid ½ min.; Ipecacuanha
Wine 4 mins.; Syrup 12 mins.; Dill Water 1
drm. Dose 1 drm. *St. Mary's.*

Mistura Potassæ Chloratis.

Chlorate of Potash 3 grs.; water 1 drm.
Chlorate of Potash 1 gr.; Dil. Hydrochloric Acid
1 min.; Glycerine 15 mins.; water to 1 drm.
Samaritan.

Mistura Potassæ Citratis. (*See also* p. 143.)

Citrate of Potash 2½ grs.; water 1 drm. *Children's.
Charing Cross.*
Bicarbonate of Potash 20 grs.; Citric Acid 18 grs.;
Syrup 3 drms.; water to 1½ oz. Dose 1 to 2
drms. *King's.*

Mistura Potassæ Liquoris c. Ipecacuanhâ.

Solution of Potash 2 mins.; Ipecacuanha Wine 2
mins.; Syrup of Red Poppies 8 mins.; Mucilage 16 mins.; water to 1 drm. *King's.*

Mistura Potassii Bromidi.

Bromide of Potassium 3 grs.; Glycerine 15 mins.;
water to 1 drm.

Mistura Quiniæ.

Sulphate of Quinia ¼ gr.; Diluted Sulphuric Acid 1
min.; water 1 drm.

Sulphate of Quinia ½ gr.; Diluted Sulphuric Acid 1 min.; Syrup 20 mins.; water to 1 drm. *Samaritan.*

Mistura Rhei Alkalina.

Aromatic Spirit of Ammonia 3 mins.; Aromatic Powder of Chalk 5 grs.: Tincture of Rhubarb 5 mins.; Peppermint water to 1 drm. *Charing Cross.*

Mistura Rhei c. Magnesiâ.

Rhubarb 7 grs.; Carbonate of Magnesia 10 grs.; Peppermint water to 1 oz. *Royal Free.*

Mistura Rhei c. Magnesiæ Sulphate.

Sulphate of Magnesia 5 grs.; Tincture of Rhubarb 10 mins.; Syrup of Ginger 5 mins.; Caraway water to 1 drm. *Children's. King's.*

Mistura Rhei c. Sodâ.

Powder of Rhubarb with Soda 1½ gr.; Syrup of Ginger 5 mins.; Peppermint Water to 1 drm.

Mistura Salina c. Ipecacuanhâ.

Solution of Citrate of Ammonia 1½ oz.; Ipecacuanha Wine 40 mins.; Antimonial Wine 30 mins. Dose 1 drm. *King's.*

Mistura Scillæ Composita.

Oxymel of Squill 5 mins.; Comp. Tincture of Camphor 2 mins.; Spirit of Nitrous Ether 2½ mins.; water to 1 drm. *Charing Cross.*

Mistura Sedativa.

Chloral Hydrate 1 gr.; Tincture of Belladonna 1 min.; Glycerine 30 mins.; water 30 mins.
Bicarbonate of Soda 2 grs.; Tincture of Opium ½ min.; Syrup 10 mins.; Dill Water 1 drm. *Samaritan.*

Mistura Senegæ.

Carbonate of Ammonia 1 gr.; Spirit of Chloroform 2 mins.; Syrup of Tolu 10 mins.; Infusion of Senega 1 drm. *Samaritan.*

Mistura Senegæ c. Ammoniâ.
> Carbonate of Ammonia ½ gr.; Spirit of Chloroform 1½ min.; Syrup 15 mins.; Infusion of Senega 30 mins.; water to 1 drm.

Mistura Sennæ c. Scillâ.
> Syrup of Senna 20 mins.; Syrup of Poppies 20 mins.; Oxymel of Squill 20 mins. *Samaritan.*

Mistura Zinci Oxidi.
> Oxide of Zinc 1 gr.; Mucilage 10 mins.; Dill water to 1 drm. *London.*

OLEUM SULPHURIS.
> Precipitated Sulphur 150 grs.; Olive Oil to 1 fl. oz.

PASTA AMYLI IODIDI.
> Starch 1 oz.; Glycerine 2 oz.; water 6 oz.: boil together, and when nearly cold add Solution of Iodine B. P. 1 oz.

PIGMENTUM IODI c. PICE.
> Iodine 120 grs.; Rectified Oil of Tar 1 oz.

PULVERES.

Pulvis Alterativus. (*See also* p. 205.)
> Grey Powder 1 gr.; Compound Cinnamon Powder 1 gr.; Rhubarb 2 grs.; Carbonate of Magnesia 2 grs. Dose 6 to 12 grs. *Royal Free.*

Pulvis Bismuthi.
> Carbonate of Magnesia 1½ gr.; Nitrate of Bismuth 1½ gr.; Aromatic Powder of Chalk 1½ gr.

Pulvis Bismuthi Astringens.
> Subnitrate of Bismuth 2 grs.; Tannic Acid 1 gr.; Aromatic Powder of Chalk 1 gr.: for a dose. *London.*

Pulvis Bismuthi Comp.
Subnitrate of Bismuth 2 grs.; Carbonate of Magnesia 1½ gr.; Aromatic Powder of Chalk 1½ gr.: for a dose. *London.*

Pulvis Calcii Sulphidi.
Sulphide of Calcium ⅛ gr.; Sugar of Milk to 1 gr.

Pulvis Catharticus.
Jalap 5 grs.; Rhubarb 5 grs.; Scammony 5 grs.: Ginger 2 grs.; Grey Powder 3 grs. Dose 5 to 15 grs. *Royal Free.*

Pulvis Cinchonæ et Sodæ.
Powder of Bark 5 grs.; Bicarbonate of Soda 5 grs. *King's. Middlesex.*

Pulvis Hydrargyri c. Cretâ Aromaticus.
Mercury with Chalk 1 gr.; Comp. Powder of Mercury with Chalk 1 gr.; Aromatic Powder of Chalk 1 gr.
Grey Powder 1½ gr.; Dover's Powder ½ gr.; Aromatic Powder of Chalk 1 gr. *London.*

Pulvis Hydrargyri c. Cretâ Compositus.
Grey Powder ½ gr.; Dover's Powder ½ gr.
Grey Powder 1½ gr.; Dover's Powder ½ gr.: for a dose. *London.*

Pulvis Hydrargyri c. Cretâ et Doveri.
Grey Powder and Comp. Powder of Ipecacuanha equal parts. *Royal Free.*

Pulvis Hydrargyri c. Cretâ et Rheo.
Grey Powder 1 gr.; dried Carbonate of Soda 1 gr.; Ginger ½ gr.; Rhubarb 2½ grs.: for a dose. *London.*

Pulvis Hydrargyri c. Cretâ c. Saccharo.
See page 209.

Pulvis Hydrargyri c. Cretâ et Sodâ.
Grey Powder 1 gr.; Bicarbonate of Soda 2 grs. *London. Royal Free.*

Pulvis Hydrargyri c. Magnesiâ.
Grey Powder 1 gr.; Rhubarb 1 gr.; Cinnamon 1 gr.; Light Carbonate of Magnesia 2 grs. *Middlesex.*

Pulvis Hydrargyri Opiatus.
Mercury with Chalk 1½ gr.; Dover's Powder ½ gr. *St. Mary's.*

Pulvis Hydrargyri et Rhei.
Mercury with Chalk 1 gr.; Rhubarb 2 grs. *St. Mary's.*
Mercury with Chalk 1 gr.; Bicarbonate of Soda 5 grs.; Rhubarb 5 grs.; Ipecacuanha ⅙ gr. *Charing Cross.*

Pulvis Hydrargyri c. Sodâ.
Mercury with Chalk 2½ grs.; Bicarbonate of Soda 2½ grs. *St. Mary's.*

Pulvis Ipecacuanhæ Comp. c. Rheo.
Dover's Powder 2 grs.; Rhubarb 3 grs. *King's.*

Pulvis Jalapæ c. Hydrargyro c. Cretâ.
Grey Powder 1 gr.; Comp. Powder of Jalap 2 grs.

Pulvis Jalapæ c. Hydrargyri Subchlorido.
Calomel ⅓ gr.; Compound Powder of Jalap 3 grs.

Pulvis Lycopodii c. Acido Carbolico.
Carbolic Acid (No. 2 Calvert) 15 grs.; Oil of Thyme 3 mins.; Rectified Spirit 10 mins.: dissolve and add of Lycopodium 300 grs. Mix well for local application. *London.*

Pulvis Rhei c. Calomelane.
Rhubarb 4 grs.; Calomel 1 gr.; Aromatic Powder of Chalk 1 gr. *Middlesex.*

Pulvis Rhei c. Hydrargyro.
Rhubarb 6 grs.; Grey Powder 3 grs. *University.* 2 to 10 years.
Rhubarb 3 grs.; Grey Powder 2 grs. *Middlesex.*

Pulvis Rhei c. Hydrargyro c. Cretâ.
Grey Powder 1 gr.; Cinnamon Powder ½ gr.; Rhubarb 3 grs.

Pulvis Rhei c. Jalapâ.
Rhubarb 1 gr.; Compound Jalap Powder 2 grs. *Middlesex.*

Pulvis Rhei et Potassæ.
Rhubarb 4 grs.; Sulphate of Potash 6 grs. *Samaritan.*

Pulvis Rhei c. Sodâ. (*See also* p. 213.)
Rhubarb 3 grs.; Carbonate of Soda 1½ gr.; Ginger ½ gr.
Rhubarb 2 grs.; Bicarbonate of Soda 4 grs. *Royal Free.*
Rhubarb 3 grs.; dried Carbonate of Soda 1 gr.: for a dose. *London.*
Rhubarb 6 grs.; dried Carbonate of Soda 3 grs.: mix. *University.* 2 to 10 years.
Rhubarb 6 grs.; Bicarbonate of Soda 3 grs.; Ginger 1 gr. *Middlesex.*
Rhubarb 6 grs.; Bicarbonate of Soda 2 grs.; Ginger 1 gr. Dose 5 to 10 grs. *St. Mary's.*

Pulvis Santonini.
See Pulvis Vermifugus.

Pulvis Santonini c. Hydrarg. Subchlor.
Santonine 3 grs.; Compound Powder of Scammony 3 grs.; Calomel ½ gr. *St. Mary's.*
Santonine 1 gr.; Calomel ½ gr.; Sugar 1½ gr.: for a dose. *London.*

Pulvis Santonini Comp.
Santonine 1 gr.; Comp. Powder of Scammony 2½ grs.; Calomel ½ gr.: for a dose. *London.*
Santonine 3 grs.; Comp. Powder of Scammony 10 grs. *Charing Cross.*

Pulvis Scammonii c. Calomelane.
Scammony 4 grs.; Calomel 2 grs.; Sugar 2 grs. *Middlesex.*

Pulvis Scammonii et Calomelanos c. Jalapâ.
Calomel 1 gr.; Scammony 2 grs.; Jalap 4 grs. *Middlesex.*

Pulvis Scammoniæ c. Hydrargyri Subchlorido.
Comp. Powder of Scammony 2 grs.; Calomel ½ gr.

Pulvis Sodæ c. Hydrargyro. (*See also* p. 215.)
Grey Powder ½ gr.; Bicarbonate of Soda 1 gr.;
Aromatic Powder of Chalk 2 grs.
Grey Powder 2 grs.; Bicarbonate of Soda 4 grs.;
Aromatic Powder of Chalk 4 grs. *Middlesex.*

Pulvis Vermifugus vel Santonini.
Santonine 3 grs.; Powder of Scammony with Subchloride of Mercury 4 grs. *Children's.* This powder is for children of 3 years old and upwards.
Santonine 2 grs.; Sugar 3 grs. *Gt. Northern.*
Santonine 3 grs.; Sugar 12 grs. *University.* 2 to 10 years.
Santonine 3 grs.; Sugar 2 grs. For children from 1 to 2 years. *Samaritan.*

Pulvis Zinci et Amyli
Oxide of Zinc 1 part; Starch 2 parts. *University. London.*

Pulvis Zinci Oxidi.
See Dusting Powder. p. 254.

UNGUENTA.

Unguentum Acidi Boracici.
Boracic Acid 1.; White Wax 1; Lard 2; Oil of Almonds 2 parts.

Unguentum Acidi Chrysophanici.
Chrysophanic Acid 10 grs.; Lard 1 oz.

Unguentum Diachyli.
Litharge 3; Olive Oil 4; Lard 2; Yellow Wax 1 part.

Unguent. Hydrarg. Oxid. Flav.
Yellow Oxide of Mercury 8 grs.; Vaseline 1 oz.

Unguentum Rubrum Mitius.
Red Oxide of Mercury 5 grs.; Sulphide of Mercury 5 grs.; Creasote 1½ min.; Lard 1 oz.

Unguentum Simplex.
Yellow Wax ¼ oz.; Olive Oil ½ oz.; Lard ¾ oz.

Unguentum Staphisagriæ.
Oil of Stavesacre 60 grs.; Lard 1 oz.

Unguentum Sulphuris Compositum.
Sulphur ⅜ oz.; Ammoniated Mercury 12 grs.;
Sulphide of Mercury 12 grs.; Olive Oil 1½ drm;
Creasote 1 min.; Lard ¾ oz.

Unguentum Sulphuris c. Zinco.
Sulphur Ointment, Zinc Ointment, equal parts.

Unguentum Viride.
Ointment of Green Elder Leaves 2 oz.; Copaiba 3 oz.; Ointment of Elemi 16 oz.

VAPOR.

Vapor Acidi Carbolici Comp.
Fir Wood Oil 4 drms.; Carbolic Acid 6 drms.;
Juniper Oil 2 drms.; Comp. Tincture of
Benzoin 1 oz; water to 10 oz.
For Inhalation, add 1 part of the solution to 6 parts of hot water.

VINUM.

Vinum Ferri.
Ammonio-Citrate of Iron 5 oz.; Rectified Spirit 40 oz.; water to 2 gallons. 1 drm. for a dose.
Ammonio-Citrate of Iron 2 grs; Rectified Spirit 4 mins.; Tincture of Orange 4 mins.; Water to 1 drm. *Royal Free.*
Tartarated Iron 1 gr.; Rectified Spirit 7½ mins.; water to 1 drm. *Charing Cross.*

HOW TO BRING UP CHILDREN.

1. Keep them warm. Let the clothing be warm and not tight. Children should wear stockings nearly up to the knees, and frocks well up to the neck, with long sleeves. Give them plenty of fresh air; take them out whenever the weather is fine. Wash the child all over with warm water daily. If possible, let the child sleep in a cot by itself. Open the windows at least twice a day.

Food.

2. If the mother has plenty of breast-milk, the child should not have any other food whatever, till it is seven months old. It must be suckled every two hours during the first month. Too frequent suckling is a common cause of the sickness of infants.

3. If the mother has only a little milk, let the child have it; and in addition, milk mixed as directed in Rule 4. Except when ordered by a doctor, do not, on any account, give the child any baked flour, arrowroot, cornflour, biscuits, tops and bottoms, or any so-called "infants' food," before it is seven months old. It is a good plan, if the mother has only a little breast-milk, that she should drink a tea-cupful of cow's milk half-an-hour before suckling.

4. If the child must be brought up entirely by hand, it should be fed with warm milk and water out of a bottle. As a rule in towns, condensed milk suits better than cow's milk, especially in summer. Be careful not to use the condensed milk too strong. Not more than one teaspoonful should be used to four or five tablespoonfuls of water. If the baby is under a month old, half-a-tea-spoonful of condensed milk will be enough. No sugar is needed. If fresh cow's milk be used, give equal parts of milk and water during the first month, and afterwards, twice as much milk as water. Add a small lump of white sugar to each pint. If milk and water is not well digested, try barley water instead of simple water. The child should be fed every two hours during the first month; every three hours afterwards. The best kind of bottle is the old-fashioned straight one, with an india-rubber teat, and without any tube at all. It is almost impossible to prevent the corks and tubes of the new kind of bottles from becoming foul, and turning the milk sour. When the child has finished its meal, the bottle must be rinsed, and placed in clean

water until again required. Once a day it should bo carefully cleaned with hot water, in which a little soda has been dissolved, and then rinsed with water and a few tea-leaves. On no account must the milk that remains after the child has finished, be used again. A fresh quantity ought to be prepared.

5. When the child has reached tho age of seven months, it should have one or two meals a day of milk thickened with Chapman's entire wheaten flour. Make a tea-spoonful of flour into a paste with cold water, pour half-a-pint of boiling water on it, and let it simmer for twenty minutes, constantly stirring; then add it to tho milk, which must be warmed. The other meals should be of milk only. Instead of Chapman's flour—Mellin's food, Savory and Moore's food, Ridge's food, or powdered Robb's biscuits, may be used. Nestle's food is made with water only, and may bo used with advantage at the age of seven months. The amount of these foods used may be increased as the child grows older. On no account keep the baby at the breast after it is twelve months old. This is injurious both to mother and child. At eight months, the child may have broth or beef tea, in addition to the milk. At twelve months, give an egg or a little milk pudding. At eighteen months, give a little meat every day, scraped or pounded into soft pulp; or a little mashed potato, with gravy. Cheese, beer, spirits, should never be given to children. It is a great mistake to give children under two years old, "just what the parents have." It is a great mistake to give them tea, instead of milk. On no account give babies teething powders and soothing syrups; Godfrey's cordial, Stedman's powders, Daffy's elixir, and Winslow's soothing syrup ought not to be given.

MEDICAL COMMITTEE.
July, 1880.

DIET TABLES.

CHARING CROSS HOSPITAL.

No. 1 Diet.*

Milk 4 pints; Beef Tea 1 pint.

No. 2 Diet.*

Breakfast.—Milk 1 pint; Bread 3 oz.; Butter ½ oz.
Dinner.—Broth 1 pint, or Beef Tea 1 pint; Bread 3 oz.; Milk Pudding.
Supper.—Milk 1 pint; Bread 3 oz.; Butter ½ oz.

No. 3 Diet.*

Breakfast.—Milk ½ pint; Bread 4 oz.; Butter ½ oz.
Dinner.—Meat (cooked) 4 oz.; Bread 4 oz.; Potatoes ½ lb.; Milk Pudding.
Supper.—Milk ½ pint; Bread 4 oz.; Butter ½ oz.

No. 4 Diet.*

Breakfast.—Milk ½ pint; Bread 4 oz.; Butter ½ oz.
Dinner.—Meat (cooked) 6 oz.; Bread 4 oz.; Potatoes ½ lb.; Milk Pudding.
Supper.—Milk ½ pint; Bread 4 oz.; Butter ½ oz.

---

* Wines, Spirits, Ale, Porter, Fish, Eggs, Pudding, and other extras may be ordered for Patients on any Diet, at the discretion of the Medical Officer.

CITY OF LONDON HOSPITAL FOR DISEASES OF THE CHEST, VICTORIA PARK.

MIXED DIET.	MILK DIET.
12 oz. of Bread. ¾ oz. of Butter. Cocoa or Coffee with Milk and Sugar for Breakfast.	12 oz. of Bread. ¾ oz. of Butter.
6 oz. of Rice, Sago, Tapioca, or Bread Pudding, 6 oz. of Boiled or Roast Meat, 6 oz. of Potatoes, for Dinner.	Cocoa or Coffee with Milk and Sugar, or ½ pint of Milk, for Breakfast.
If Fish, Chop, or Poultry be ordered, Meat to be omitted.	½ pint of Milk with 12 oz. of Rice, Sago, Tapioca, or Bread Pudding, for Dinner.
Tea with Milk and Sugar for Tea. Milk or Beef Tea for Supper.	Tea with Milk and Sugar or Milk for Tea.

Children under 10 years of age to be allowed two-thirds the quantity of each Diet prescribed for adults.

The quantity of Meat sent up to each Patient to be 4 oz., but an extra allowance, equivalent to the quantity in the list, will be sent up into the Ward, to be distributed to such Patients as require it.

The quantity of Tea, Coffee, Cocoa, or Milk to be three-quarters of a pint for Men and half a pint for Women.

Wine, Spirits, Porter, Chops, Poultry, Fish, or Eggs, &c., to be served when specially ordered.

The quantity of Porter or Ale issued to each Male Patient to be three-quarters of a pint, and to each female half a pint, unless otherwise directed.

CHILDREN'S HOSPITAL

	MILK DIET.	BROTH DIET.
Breakfast. 8 o'clock.	Milk ½ pint. 2 oz. Bread with Butter.	Milk or Cocoa, ½ pint. Bread with Butter, 2 oz.
Dinner. 12 o'clock.	Rice or other Milk Pudding. Milk ⅓ pint.	Mutton Broth, made with vegetables, ⅓ pint. Bread 1 oz.
Tea. 4 o'clock.	Milk ⅓ pint, with 2 oz. of Bread and Butter, if desired.	Bread 2½ oz., with Dripping or Butter. Milk ⅓ pint.
Supper. 6 o'clock. A portion to be set aside for the night and early morning.	Milk ½ pint, with 2 oz. of Bread and Butter, if desired.	Bread 2 oz. with Butter. Milk ⅓ pint.

It is to be understood that Water is to be given with the Milk, or otherwise, as required.

With all Diets including Potatoes, Greens, Turnips, or Parsnips are to be substituted twice a week, when practicable.

EXTRAS, as Tea and Sugar, Mutton Chops, Fowl, Eggs, Strong Beef Tea, &c., may be ordered, as may Wine, Beer, or Spirit, for any of the Patients for whom the Medical Officers think fit to prescribe them.

"*Fancy Diet*" may be ordered in exceptional cases. Under this the child is allowed whatever he can take—meat, fish, fowl, sausage, &c., with frequent variation.

RECEIPTS FOR THE ABOVE DIETS.

Cocoa.—¼ lb. Cocoa Nibs to each gallon of Water, set on the hot plate all day, strain next morning and serve, ¼ Pint of Milk added to each Pint of Cocoa, and ¼ oz. Sugar.

Rice Puddings.—1 oz. Rice to each Pint of Milk, placed in the oven for three hours.

CHILDREN'S HOSPITAL.

Beef Tea Diet.	Fish Diet.	Meat Diet.
Milk ½ pint. 2 oz. Bread, with Butter.	Milk or Cocoa, ½ pint. Bread 2½ oz. with Butter.	Milk or Cocoa, ½ pint. Bread 2¼ oz. with Butter.
½ pint Beef Tea. 1 oz. Bread.	Sole 2½ oz. boiled. Bread 1 oz. Mashed Potatoes, 3 oz.	Roast or Boiled Mutton, or Roast Beef, 2½ oz. Mashed Potatoes 4 oz.
2 oz. Bread and Butter. Milk ½ pint.	Bread 2½ oz. with Dripping, Butter, or Treacle. Milk ¼ pint.	Bread 2 oz., with Dripping, Butter, or Treacle. Milk ¼ pint.
½ pint Beef Tea. ½ pint of Milk for the night.	Bread 2 oz., with Butter or Dripping. Milk ¼ pint.	Bread 2 oz., with Butter or Dripping. Milk ¼ pint.

Custard Puddings.—1 Egg to ½ Pint Milk.

Mashed Potatoes.—Plainly boiled. Salt added and then mashed, without Milk.

Mutton Broth, with Meat.—1 lb. Neck of Mutton to each Pint Cold Water, add Carrots, Turnips, Onions, and Barley—let it all simmer together for 3 hours, and serve.

Mutton Broth, without Meat —2 "Shank ends" to a Pint of Cold Water, simmer for 3 hours and strain. Serve with Vegetables as the other.

Beef Tea.—Cut up very small ¾ lb. lean Beef, place the meat in 1½ pint Cold Water, to stand one hour; add ¼ lb. good Stock Meat, place on the stove to simmer slowly for 12 hours. The liquid, now reduced to 1 Pint, to be then poured off; when cold skim carefully.

Strong Beef Tea.—Same as above, doubling the proportion of Meat and Stock.

All Diets are adjusted for the age of 7 years, the mean age of the children in the Hospital; the apportionment is to be arranged in the Ward according to the age and needs of each child.

ROYAL HOSPITAL FOR DISEASES OF THE CHEST.

ALLOWANCE FOR THE WEEK FOR EACH PATIENT.
Tea 2 oz.; Butter 6 oz.; Sugar 8 oz.

DAILY ALLOWANCE.

ORDINARY DIET (in addition to the above).	MILK DIET (in addition to the above).
Breakfast, 8 A.M. :— Bread 12 oz. (for the day). Milk or Cocoa ½ pint. *Dinner*, 12 Noon ;— Meat (cooked) 4 oz.* Potatoes 8 oz. Rice, Tapioca, Sago, or Bread Pudding, 6 oz. (made with Milk and Eggs). *Tea*, 4 P.M. :— Milk ¼ pint. *Supper*, 8 P.M. :— Milk ½ pint.	The Day's Allowance :— Bread 8 oz. Milk 1¾ pint. Cocoa ½ pint. Bread Pudding 8 oz. or Rice Pudding 8 oz.

* If desired by the Patient an extra allowance of Meat up to 6 oz. will be sent up. Every Patient on being admitted into the Hospital to be placed on Milk Diet until further orders.

EXTRAS (not to be placed on the Diet List without the authority of the Physician, except in urgent cases, when the House Physician may order them), Mutton Chops, Fish, Poultry, Eggs, Beef Tea (¾ lb. to the pint), Porter, Ale, Wines (Port, Sherry, or Claret), Spirits (Brandy, Whiskey, Gin, Rum), Lemonade, Soda Water, &c.

The Joints to be :—Sunday, Roast Beef; Monday, Roast Mutton; Tuesday, Boiled Mutton; Wednesday, Roast Beef; Thursday, Roast Mutton; Friday, Boiled Mutton; Saturday, Roast Mutton.

HOSPITAL FOR CONSUMPTION AND DISEASES OF THE CHEST, BROMPTON.

GENERAL DAILY DIET TABLE.

	Breakfast, 8.30 a.m.	Dinner, 12.30 or 1 p.m.	Tea, 5 p.m.	Supper, 8 p.m.
Sunday	Bread and Butter. Egg or Bacon. Coffee or Cocoa.	Fresh Beef, Roast or Boiled. Potatoes and Parsnips or Carrots. Rice Pudding.	Bread. Butter or Treacle. Tea or Cocoa.	Milk, or Gruel, or Soup, with Vegetables.
Monday......	Do.	Roast Leg of Mutton. Potatoes. Macaroni Pudding.	Do.	Do. or Barley Soup.
Tuesday......	Do.	Roast Leg of Mutton. Potatoes and Greens. Sago Pudding.	Do.	Do. or Rice Soup.
Wednesday	Do.	Boiled Leg of Mutton. Potatoes and Turnips. Tapioca Pudding.	Do.	Do. or Macaroni Soup.
Thursday ...	Do.	Roast Leg of Mutton. Potatoes and Greens. Macaroni Pudding.	Do.	Do. or Sago Soup.
Friday	Do.	Roast Leg of Mutton. Potatoes and Lentils. Rice Pudding.	Do.	Do. or Soup with Vegetables.
Saturday ...	Do.	Roast Leg of Mutton. Potatoes. Sago Pudding	Do.	Do. or Rice Soup.

HOSPITAL FOR CONSUMPTION AND DISEASES OF THE CHEST, BROMPTON—*continued*.

GENERAL DIET.

CONSTITUENTS.	QUANTITY. MEN. WOMEN.	DAYS.
Meat { Silver side of Round of Beef, or Leg of Mutton }	6 oz. 4 oz. cooked.	} Daily.
Bread	13 oz.	,,
Butter	1 oz.	,,
Potatoes	5½ oz.	,,
Greens, when in season..		Twice a week.
Carrots, do.		Once a week.
Turnips, do.		,,
Parsnips, do.		,,
Lentils, do.		,,
Egg or Bacon		Daily.
Coffee or Cocoa	½ pint	,, A.M.
Tea or Cocoa	do.	,, P.M.
Pudding	8 oz.	,,
Milk, Gruel, or Soup	½ pint	,,
Treacle		,,

EXTRAS.

(Not to be placed on the Diet List without the authority of the Physician, except in urgent cases when the Resident Medical Officer may order them.)

Chop—Tuesday, Saturday, and Sunday.

Fish—¼ lb. Sole, Whiting, &c. (Fried or Steamed), Monday, Wednesday, and Friday.

Oysters, Eels Stewed. Minced and Sieved Meat, Rabbit, Fowl. Strong Beef Tea, Mutton Broth, Veal Tea. Milk, Suet, Arrowroot. Porter, Ale. Wines—Sherry, Port, Claret. Spirits—Brandy, Gin, Whiskey, Rum. Lemonade, Soda Water, Ginger Beer.

LONDON FEVER HOSPITAL.

MEN.

Low Diet.
Bread 4 oz. ; Milk ½ pint ; Gruel 1 pint ; Sugar ¼ oz.

Beef Tea Diet.
Beef Tea 1 pint ; Milk 1 pint ; Bread 4 oz.

Middle Diet.
Bread 10 oz. ; Broth 1 pint ; Milk 1 pint ; Rice or Bread (for pudding) 2 oz. ; Egg (for pudding) 1 ; Sugar (for pudding) ½ oz.

Fish Diet.
Bread 12 oz. ; Fish (sole, haddock, cod, or brill, uncooked) 8 oz. ; Potatoes 8 oz. ; Cocoa 1 oz. ; Sugar ½ oz. ; Milk ⅙ pint.

Full Diet.
Bread 16 oz. ; Meat (uncooked and without bone) 12 oz. ; Potatoes 12 oz. ; Cocoa 1 oz. ; Sugar ½ oz. ; Milk ¼ pint ; Beer 1 pint.

Extras.
Beef Tea, Strong Beef Tea, and Eggs, as ordered ; Arrowroot ½ oz. ; Custard Pudding—1 egg, ½ pint Milk, ½ oz. Sugar ; Tea ¼ oz. per day ; Sugar 1 oz. per day Butter 1 oz. per day.

LONDON FEVER HOSPITAL—*continued*.

WOMEN.

LOW DIET.
Bread 4 oz. ; Milk ½ pint ; Gruel 1 pint ; Sugar ¼ oz.

BEEF TEA DIET.
Beef Tea 1 pint ; Milk 1 pint ; Bread 4 oz.

MIDDLE DIET.
Bread 8 oz. ; Broth 1 pint ; Milk 1 pint ; Rice or Bread (for pudding) 2 oz. ; Egg (for pudding) 1 ; Sugar (for pudding) ½ oz.

FISH DIET.
Bread 10 oz. ; Fish (sole, haddock, cod, or brill, uncooked) 8 oz. ; Potatoes 8 oz. ; Cocoa 1 oz. ; Sugar ½ oz. ; Milk ⅛ pint.

FULL DIET.
Bread 12 oz. ; Meat (uncooked and without bone) 10 oz. ; Potatoes 12 oz. ; Cocoa 1 oz. ; Sugar ½ oz. : Milk ¼ pint ; Beer ½ pint.

EXTRAS.
Beef Tea, Strong Beef Tea, and Eggs, as ordered ; Arrowroot ½ oz. ; Custard Pudding—1 Egg, ¼ pint Milk, ½ oz. Sugar ; Tea ¼ oz. per day ; Sugar 1 oz. per day ; Butter 1 oz. per day.

ROYAL FREE HOSPITAL.

No. 1.
Coffee ¾ pint; Milk 2 oz. morning and evening.
Beef Tea ½ pint; Gruel 1 pint.; Bread 8 oz.

No. 2.
8 o'clock, ¾ pint Coffee, with 2 oz. of Milk.
12 ,, 1 pint of Soup, with 3 oz. of Boiled Rice.
5 ,, ¾ pint of Coffee, with 2 oz. of Milk.
8 ,, 1 pint of Gruel.

No. 3.
8 o'clock, ¾ pint of Coffee, with 2 oz. of Milk.
12 ,, ½ lb. Meat, uncooked } Males.
½ lb. Potatoes
6 oz. Meat, uncooked } Females.
½ lb. Potatoes
5 ,, ¾ pint of Coffee, with 2 oz. of Milk.
8 ,, 1 pint of Gruel.
Bread 12 oz. to each Patient on Diet No. 2 or 3.

EXTRAS that may be ordered with all Diets:—Wine, Brandy, Gin, Porter, Ale, Lemonade, Soda Water, Tea instead of Coffee if the Patients cannot find it for themselves.

EXTRAS that may be ordered with No. 1 and No. 2 Diets only:—Chops, Fish, Slices off Joint, Eggs, Fresh Greens, Beef Tea, Soup, Rice and Batter Pudding, Arrowroot, Milk.

DAILY MEAT BILL OF FARE.
Monday—Mutton Roast. | Thursday—Mutton Boiled.
Tuesday—Mutton Boiled. | Friday—Mutton Roast.
Wednesday—Beef Roast. | Saturday—Mutton Boiled.
Sunday—Beef Roast.

The Soup to be made with Mutton and the liquor from the Boiled Joints, with Onions, Turnips, and Rice.

All new Patients to be put on No. 1 or No. 2 Diet till seen by their Medical Officer. No order for Extras to run for more than one day unless renewed.

GREAT NORTHERN HOPITAL.

Low Diet.

Tea or Coffee 1 pint; Milk 1½ pint; Bread 12 oz.; Broth ½ pint; Barley Water *ad libitum*; Arrowroot, Sago, or Rice if specially ordered.

Milk Diet.

Milk 3 pints; Barley Water *ad libitum*. Bread 12 oz.; Rice Pudding ½ lb. if specially ordered.

Middle Diet.

Breakfast.—Tea or Coffee ¾ pint; Milk ¼ pint; Bread 12 oz.; Butter ½ oz.
Dinner.—Meat 4 oz.; Broth 1 pint; Potatoes, 8 oz.
Tea.—Tea or Coffee ¾ pint; Milk ¼ pint.
Supper.—Broth ½ pint.

Full Diet.

Breakfast.—Tea or Coffee ¾ pint.; Milk ¼ pint; Bread 12 oz.; Butter ½ oz.
Dinner.—Meat 6 oz. M., 4 oz. F.; Broth 1 pint; Potatoes 8 oz.
Tea.—Tea or Coffee ¾ pint; Milk ¼ pint.
Supper.—Broth 1 pint.
Beef two days in the week, Mutton the other days.

Extras.

Mutton Chops, Beef Tea, Fish, Eggs, Green Vegetables, Extra Bread, Extra Milk, Beer, Porter, Wine, Brandy, Gin.

Broth.—1 lb. of Meat to a gallon of Pot Liquor, thickened with Rice, Pearl Barley, or Oatmeal, and 1 oz. Onions.

Beef Tea.—Meat 1 lb. to Water 1 pint.

Gruel.—Grits ½ oz., Water ½ pint, Milk ⅓ pint, Sugar ½ oz.

Rice Pudding.—Carolina Rice ½ oz., Sugar ¼ oz., Milk ⅓ pint.

GUY'S HOSPITAL.

Full Diet.	Middle Diet.
Fourteen Ounces of Bread. One Ounce of Butter. One Pint of Porter for Males. Half-pint of Porter for Females. Milk may be substituted. Six Ounces of Dressed Meat, roasted and boiled, alternately, with half-pound of Potatoes. Half a pint of Mutton Broth, in addition, on days when boiled Meat is given. Half-pound of Baked Rice Pudding, made with Milk on Roast Meat days.	Twelve Ounces of Bread. One Ounce of Butter. Half-pint of Porter for Males and Females. Milk may be substituted. Four Ounces of Dressed Meat, roasted and boiled alternately, with half-pound of Potatoes. Half a pint of Mutton Broth, in addition, on days when boiled meat is given. Half-pound of Baked Rice Pudding made with Milk on Roast Meat days.

Tea and Sugar morning and evening with all Diets.

Low Diet.	
Ten Ounces of Bread. One Ounce of Butter. Chop or Fish with half-pound of Potatoes. Beef Tea. Eggs. Milk.	Extra Milk and Eggs may be ordered with Full or Middle Diet. Chicken and other delicacies will be given with the Low Diet when specially ordered. Porridge, Gruel and Barley Water are given with all Diets as required. Wines and Spirits if continued must be mentioned each time the Physician or Surgeon attends.

Tea and Sugar morning and evening with all Diets.

KING'S COLLEGE HOSPITAL.

MEN.

DAILY ALLOWANCE.

FULL DIET.

Breakfast.—Bread 6 oz.; Milk ¼ pint.
Dinner.—Meat (cooked) 6 oz.; Bread 6 oz.; Potatoes ½ lb.; Porter 1 pint.
Supper.—Gruel 1 pint; Milk ¼ pint.

MIDDLE DIET.

Breakfast.—Bread 6 oz.; Milk ¼ pint.
Dinner.—Meat (cooked) 4 oz.; Bread 6 oz.; Potatoes ½ lb.; Porter ½ pint.
Supper.—Gruel 1 pint : Milk ¼ pint.

MILK DIET.

Breakfast.—Bread 4 oz.; Milk ½ pint.
Dinner.—Bread 4 oz.; Rice Milk ½ pint (four days); Rice or Bread Pudding ½ lb. (three days).
Supper.—Milk ½ pint.

CHILDREN'S DIETS (under ten years of age)—two-thirds of any Diet ordered.
Roast Mutton—Monday and Thursday.
Boiled Mutton—Tuesday and Friday.
Stewed Mutton—Wednesday and Saturday.
On Sundays—Roast Beef.

DIET TABLES.

King's College Hospital—*continued*

WOMEN.

DAILY ALLOWANCE.

FULL DIET.

Breakfast.—Bread 6 oz. ; Milk ¼ pint.
Dinner.—Meat (cooked) 4 oz. ; Bread 6 oz. ; Potatoes ½ lb. ; Porter ½ pint.
Supper.—Gruel 1 pint ; Milk ¼ pint.

MIDDLE DIET.

Breakfast.—Bread 6 oz. ; Milk ¼ pint
Dinner.—Meat (cooked) 3 oz. ; Bread 6 oz. ; Potatoes ½ lb. ; Porter ½ pint.
Supper.—Gruel 1 pint ; Milk ¼ pint.

MILK DIET.

Breakfast.—Bread 4 oz. ; Milk ¼ pint.
Dinner.—Bread 4 oz. ; Rice Milk ½ pint (four days) ; Rice or Bread Pudding ½ lb. (three days).
Supper.—Milk ½ pint.

No extras (except Wine and Spirits) to be supplied by the Steward, unless authorised by the signature of the Visiting Physician or Surgeon.

No extras allowed on Full Diet.

In any Diet, Rice or Bread Pudding may be substituted for Meat if desired.

No Patient on being admitted into the Hospital to be placed on Full Diet until ordered by the Visiting Physician or Surgeon.

LONDON HOSPITAL.

ADMISSION DIET.*
12 oz. Bread.
2 pints Milk.
1 pint Beef Tea.

(FOR CHILDREN.)
8 oz. Bread.
1 pint Milk.
½ pint Beef Tea.

FULL DIET.
12 oz. Bread.
8 oz. Potatoes.
6 oz. Meat (Roast or Boiled Leg or Shoulder of Mutton, or Roast Beef).*
1 pint Porter ⎫
    or   ⎬ as ordered.
1 pint Milk. ⎭

MIDDLE DIET.
12 oz. Bread.
8 oz. Potatoes.
4 oz. Meat (Roast or Boiled Leg or Shoulder of Mutton, or Roast Beef).†
½ pint Porter ⎫
    or   ⎬ as ordered
½ pint Milk ⎭

FEVER DIET.
2 pints Milk.
1 pint Beef Tea.

CHILDREN'S DIET.
8 oz. Bread.
6 oz. Potatoes.
2 oz. Meat.†
1 pint Milk.

HYDROCARBON DIET.
12 oz. Bread.
4 oz. Fat Bacon.†
1 pint Milk.
Pudding (1 oz. Arrowroot, Yolk of 2 Eggs, 1 pint Milk).

---

* Patients on being admitted, are to be put on "Admission Diet," unless otherwise ordered.

† NOTE.—The meat is weighed when cooked; full allowance being made for bone.

DIET TABLES.

DIABETIC DIET.

6 oz. Gluten Bread
6 oz. Meat (Roast or Boiled Leg or Shoulder of Mutton, or Roast Beef).*
Gluten Bread Pudding.†
Watercress.

SPECIAL DIETS.

Mutton Chops } (weight 8 oz. { In each case with 12 oz.
Beef Steaks } uncooked). { Bread, 8 oz. Pota-
Fish (10 oz. uncooked). { toes, and 1 pint Milk
{ or Porter as ordered.

EXTRAS.

Mutton Chops }
Beef Steaks } when specially ordered, in addition to
Fish } particular Diets.

Beef Tea   8 oz. Meat to the pint.
Special Beef Tea 16 oz.   do.
Mutton Broth   10 oz.   do.
Veal   do.   10 oz.   do.

‡Puddings { Rice }
          { Light } alternately.
          { Batter }
          { Suet, as ordered.

Eggs.   Spirits.
Rusks.   Lemonade.
Bread   } in addition to Diet   Aërated Water.
Porter } quantities.   Coffee.
Milk   }   Cocoa.
Green Vegetables.   Gruel.
Watercress.   Oatmeal Porridge.
Wines.   Arrowroot.

---

* NOTE.—The meat is weighed when cooked; full allowance being made for bone.

† *Recipe for Gluten Bread Pudding*—
   Soak 1 oz. Gluten Bread in ½ pint Milk for an hour, beat up with an Egg, and 1 oz. Gluten Flour, then put the Mixture into a mould and bake.

‡ *Recipes for Puddings* (*the quantities being sufficient for Four Patients*)—Rice.—4 oz. Rice, 2 oz. Sugar.
   Light.—6 Eggs, 2 oz. Sugar, 1½ oz. Flour.
   Batter.—4 Eggs, 2 oz. Sugar, 6 oz. Flour.
   Milk (in each case) sufficient to make up one quart of the Mixture.
   Suet.—6 oz. Suet, 1 lb. Flour, Water sufficient to make a stiff paste.

Peptonised Foods, see p. 165.

THE MIDDLE SEX

DIET TABLE.

MEALS.	SIMPLE.	ORDINARY.	FISH.	MILK DIET.
DAILY . .	10 oz of Bread.	10 oz. of Bread.	10 oz. of Bread.	10 oz. of Bread.
BREAKFAST .	½ pint of Milk.	½ pint of Milk.	½ pint of Milk.	½ pint of Milk.
DINNER . .	1 pint of strong Beef Tea.	6 oz. of undressed Meat (Leg and Shoulder of Mutton only) weighed with the bone before it is dressed—roast and boiled alternately; and ½ lb. Potatoes.	8 oz. of Fish (Whiting, Cod. Plaice, or Brill); and ½ lb of Potatoes.	ALTERNATE DAYS. Pudding. Egg. Sugar. 1½ oz. Rice... ½......½ oz. 1½ oz. Sago...½......½ oz. EXTRA. Eggs. Sugar. Custard ...2...... ..½ oz.
SUPPER . .	½ pint of Milk.	½ pint of strong Beef Tea.	½ pint of Milk.	½ pint of Milk.
EXTRAS . .	Chops, ½ lb. each when trimmed (Mutton or Pork). Steaks, rump steak ½ lb., without bone. Chicken, a quarter. Sausages, ½ lb. Rabbit, a quarter. Tripe, ½ lb. Bacon, 3 oz.		Strong beef tea, 12 oz. of clod and sticking of Beef, without bone, to a pint. Jelly beef tea, 24 oz., ditto, ditto. Broth, without meat, ¾ lb. of Neck of Mutton, with bone, to a pint. Chicken Broth, ½ chicken to a pint.	

Every Patient admitted into the Hospital is placed upon Simple Diet, until a Diet is otherwise ordered.
Diets ordered by the Physicians and Surgeons are to be continued until changed by subsequent orders.
Extras are allowed for one day only, unless the Physician or Surgeon write the word DAILY.
On Sundays, those Patients on Meat Diet have Roast Beef.
Breakfast is served at 7, Dinner at 12, Tea at 4, and Supper at 7 o'clock.
Peptonised Foods, see p. 165.

HOSPITAL.

1883.

MUTTON BROTH.	CONVALESCENT.				
10 oz. of Bread.	10 oz. of Bread.				
½ pint of Milk.	½ pint of Milk.				
8 oz. of undressed Meat (Neck of Mutton only) weighed with the bone before it is dressed—served in one pint of Broth with Barley (2 oz.) and ½ lb. of Potatoes.	MALE.	FEMALE.	Chops Steaks Chicken Sausages Rabbits or other Meat	Carry with them the simple diet and ½ lb. of Potatoes.
	12 oz. of undressed Meat (Leg and shoulder of Mutton only, except on Sundays, when the same quantity of Roast Sirloin and best Round of Beef is issued), roast and boiled alternately. ½ lb. of Potatoes.	8 oz. of undressed Meat (Leg and Shoulder of Mutton only, except on Sundays, when the same quantity of Roast Sirloin and best Round of Beef is issued), roast and boiled alternately. ½ lb. of Potatoes.		
½ pint of Milk.	½ pint of Milk.				

Oysters.
Greens.
Suet Pudding { 1 oz. of Beef Suet. / 2 oz. of Flour. }
Custard Pudding.
Arrowroot.
Sago.
Oatmeal gruel, 3 oz. to a pint.
Eggs.

Jellies.
Porter.
Ale.
Stout.
Wine.
Spirits.
Oranges.
Lemons.
Meat Essences.

Extras not mentioned in this list can only be obtained by the order being initialled daily by the Physician or Surgeon in charge, or the Resident Medical Officer.

Saturday Diets are continued until Monday, and cannot be altered except in cases of emergency.

The above Dietary does not necessarily apply to the Patients in the Cancer Wards, who are allowed special privileges according to the exigencies of the cases.

ST. BARTHOLOMEW'S HOSPITAL.

MEALS.	FULL DIET (MEAT).	HALF DIET (MEAT).
Breakfast.	1 Pint of Tea. Bread and Butter.	1 pint of Tea. Bread and Butter.
Dinner.	Half-pound Meat when dressed. Half-pound Potatoes. Bread and Beer.	Quarter-pound Meat when dressed. Half-pound Potatoes. Bread and Beer.
Tea.	1 Pint of Tea. Bread and Butter.	1 pint of Tea. Bread and Butter.
Supper.	Bread and Butter. Beer.	Bread and Butter. Beer.
Daily Allowances to each Patient.	2 pints of Tea. 14 oz. of Bread. Half-pound Meat when dressed. Half-pound Potatoes. 2 pints Beer (*Men*). 1 pint Beer (*Women*). 1 oz. Butter.	2 pints of Tea. 12 oz. Bread. Quarter-pound Meat when dressed. Half-pound Potatoes. 1 pint Beer. ¾ oz. Butter.

Extras to be specially ordered.—Mutton Chops, Beef Steaks, Beef for Beef Tea, Fish, Eggs, Pudding, Jelly, Porter, Ale, Wine, or Spirits.

ST. BARTHOLOMEW'S HOSPITAL.

Broth Diet.	Milk Diet.	Low Diet.	
1 pint of Tea.	1 pint of Tea.	Bread.	Children under Nine Years to receive half Allowances.
1 pint and a half Broth. 6 oz. Potatoes (mashed). Bread.	1 pint and a half Milk, or 1 pint Milk with Arrowroot, Rice, or Sago. Bread.	Broth, Gruel, or Barley Water, as may be ordered.	
1 pint of Tea. Bread and Butter.	1 pint of Tea. Bread and Butter.		
Bread and Butter. Gruel.	Bread and Butter. Gruel.		
2 pints of Tea. 12 oz. Bread. 1 pint and a half Broth 6 oz. Potatoes (mashed). ¾ oz. Butter. Gruel.	2 pints of Tea, 12 oz. Bread. 1 pint and a half Milk, or 1 pint Milk with Arrowroot, Rice, or Sago. ¾ oz. Butter. Gruel.	Bread. Broth, Gruel, or Barley Water.	

Each Patient on Admission to be placed on Milk Diet until a Diet is ordered by the Physician or Surgeon.

ST. GEORGE'S HOSPITAL.

Each Patient is allowed 2 oz. of Tea weekly, 8 oz. of Sugar weekly, 12 oz. of Bread daily, 1 oz. of Butter daily, ¼ pint of Milk morning and evening.

ORDINARY DIET.

Dinner.—3 oz. of cooked Meat and ½ lb. of Potatoes. ½ pint of Porter daily to men above sixteen years of age.
Supper.—1 pint of Gruel.

EXTRA DIET.

Dinner.—6 oz. of cooked Meat and ½ lb. of Potatoes. 1 pint of Porter to men above sixteen years of age.
Supper.—1 pint of Gruel.

FISH DIET.

Dinner.—4 oz. of plain boiled white Fish (as whiting, plaice, flounders, or haddock).
Supper.—1 pint of Gruel.

BROTH DIET.

Dinner.—1 pint of Broth and 6 oz. of light Pudding.
Supper.—1 pint of Gruel.

MILK DIET.

Dinner.—Four days, 1½ pint of Rice Milk. Three days ½ lb. of Bread or Rice Pudding.
Supper.—¼ pint of Milk.
Beef Tea, Arrowroot, &c., must be specially directed.

ST. MARY'S HOSPITAL.
DIET SCALE.
Adopted by the Weekly Board of 18th May, 1883.

MEALS.	ORDINARY DIET.	BROTH DIET.	SIMPLE DIET.
Breakfast. 6 A.M.	1 pint Tea, Coffee, or Cocoa, with Sugar. Bread & Butter. ¼ pint of Milk.	1 pint Cocoa or Milk, with Sugar. Bread & Butter.	1 pint Cocoa or Milk. Bread & Butter.
Dinner. 12 noon.	4 oz Meat (cooked). ½ lb. Potatoes or other Vegetables. Bread.	½ lb. Meat (uncooked). 1 pint Broth. Bread.	Beef Tea or "Extra." Bread.
Tea. 4 P.M.	1 pint Tea, Coffee, or Cocoa, with Sugar. Bread & Butter. ¼ pint Milk.	1 pint Cocoa or Milk. Bread & Butter.	1 pint Cocoa or Milk. Bread & Butter.
Supper. 7 P.M.	1 pint Beef Tea, Milk, or Cocoa. Bread & Butter.	1 pint Beef Tea, Milk, or Cocoa. Bread & Butter.	1 pint Beef Tea, Milk, or Cocoa. Bread and Beef Tea.
Daily Allowance to each Patient.	Say— 2 pints Tea with Sugar. ½ pint Milk. 1 pint Beef Tea or Milk. 4 oz. of cooked Meat. ½ lb. Potatoes or other Vegetables. 12 oz. Bread. 1 oz. Butter.	Say— 2 pints Cocoa with Sugar. 1 pint Beef Tea or Milk. 1 pint Broth. 8 oz. Meat (before cooking). 12 oz. Bread. 1 oz. Butter.	Say— 2 pints Cocoa 1 pint Beef Tea or "Extra." 1 pint Beef Tea or Milk. 12 oz. Bread. 1 oz. Butter.

NOTE 1.—No extras are to be ordered by the Resident Medical Officer in the absence of the Physician or Surgeon, except in cases of great urgency, a Special Report of which must be made to the Physician or Surgeon at his next visit, and an entry made in a book provided for that purpose, and kept by the Sister. Such entry to be initialled by the Physician or Surgeon, if the "Extra" is approved of by him.

NOTE 2.—The Ordinary Diet for Children under 7 will be half the "Ordinary." No Tea or Coffee will be allowed to Children under 7. They will have instead Milk (extra if necessary) or Cocoa.

NOTE 3.—The Physicians or Surgeons can order 2 oz. of extra Meat for such cases as they may think fit, without an entry in the "Extra Book."

NOTE 4.—The Special Fever Diet for an Adult will consist of 2 pints each of Milk and Beef Tea in the 24 hours.

NOTE 5.—When Fish is ordered, ¼ lb. of such white fish as is supplied at general contract price will be substituted for the Meat of Ordinary Diet, or for the "Extra" of the Simple Diet.

NOTE 6.—The Steward is instructed not to issue any Extra that is not initialled in the Extra Book.

ST. THOMAS'S HOSPITAL.

Daily Allowance to each Patient on either of the following Diets:—

Full Diet.	Mixed Diet.	Milk Diet.
12 oz. of Bread.	12 oz. of Bread.	12 oz. of Bread.
¾ oz. of Butter.	¾ oz. of Butter.	¾ oz. of Butter.
¾ pint of Tea with Milk and Sugar for Breakfast.	¾ Pint of Tea with Milk and Sugar for Breakfast.	¾ Pint of Tea with Milk and Sugar for Breakfast.
The same for Tea.	The same for Tea.	The same for Tea.
4 oz. of Beef or Mutton when dressed, Roast or Boiled, alternately.	4 oz. for Men, and 3 oz. for Women, of Mutton when dressed, Roast or Boiled, alternately.	
½ lb. of Potatoes or Fresh Vegetables.	¼ lb. of Potatoes or Fresh Vegetables.	
	8 oz. of Rice or Bread Pudding alternately.	8 oz. of Rice or Bread Pudding alternately.
½ pint of Milk in the Forenoon.	½ pint of Milk.	1½ pint of Milk.
Porter, &c., if ordered.	When Fish is ordered Meat to be omitted.	

Wine, Brandy, Gin, Porter, Mutton Chops, Fish, Eggs, Beef Tea, Soda Water, Lemonade, and other extras, to be served when specially ordered, such order being renewed at each regular visit of the Physician or Surgeon.

ST. THOMAS'S HOSPITAL.

Daily Allowance to each Patient on either of the following Diets :—

Fever Diet.	Children's Diet.	
	Mixed.	Milk.
4 oz. of Bread.	12 oz. of Bread.	8 oz. of Bread.
	¾ oz. of Butter.	½ oz. of Butter.
2 Pints of Barley Water or Gruel.	½ pint of Milk for Breakfast.	½ pint of Milk for Breakfast.
2 Pints of Milk.	The same for Tea.	The same for Tea.
	2 oz. of Mutton when dressed. Roast or Boiled, alternately.	
	¼ lb. of Potatoes or Fresh Vegetables.	
	6 oz. of Rice or Bread Pudding.	6 oz. of Rice or Bread Pudding.
	½ pint of Milk.	¼ Pint of Milk.

The Children's Diets are intended for all Children under ten years of age.

N.B.—Each Patient on admission into the Hospital to be placed on Milk or Fever Diet until the proper Diet is ordered by the Physician or Surgeon.

THROAT HOSPITAL.

	FULL.	MILK. MILK, 2 PINTS.	FISH.
Breakfast. 8 A.M.	Bread 4 oz.* Butter ½ oz. Tea half a pint. Sugar 1¼ oz. Milk 2 oz.	Bread 4 oz. Butter, ½ oz. Tea half a pint. Sugar 1¼ oz.	Is the same as Full Diet, but ½ lb. of Fish in place of Meat for Dinner.
Dinner. 12.30 P.M.	Bread 4 oz. 6 oz. cooked Meat.† 6 oz. Potatoes.‡ Pudding Porter half a pt.§	‖ Bread 4 oz. Beef Tea half a pint (made from ½ lb. of beef). 2 Eggs.	
Tea. 4.30 P.M.	Bread 4 oz. Butter ½ oz. Tea half a pint. Sugar 1¼ oz. Milk 2 oz.	Bread 4 oz. Butter ½ oz. Tea half a pint. Sugar 1¼ oz.	
Supper. 8 P.M.	Arrowroot or Corn Flour, with half a pint Milk, or 1 pint Gruel, or Suet and half a pint Milk, or 4 oz. Bread and half a pint of Porter.	Arrowroot, Corn Flour, Gruel, or Rice, or 1 Egg.	

* The entire quantity of Bread, Butter, and Sugar allowed for the day is allotted each morning, and may be used at different meals, according to the inclination of the Patient. † 8 oz. *uncooked* Meat—Sunday and Wednesday, Roast Beef; Monday, Thursday, Saturday, Roast Mutton; Tuesday, Boiled Mutton; Friday, Boiled Beef. ‡ 6 oz. *uncooked* Potatoes. § When Porter is not given Barley Water is allowed. ‖ Where the Patient can take no solid food, 1 pint of Beef Tea and 1 extra Egg are allowed in place of the Bread.

THROAT HOSPITAL.

HALF-FULL.	HALF-MILK.	EXTRAS TO BE ORDERED ONLY BY THE MEDICAL OFFICERS.
Is the same as Full, but only 3 oz. of cooked Meat is given, and 8 oz. of Bread allowed.	Is the same as Milk Diet, but only 1 pint of Milk and 1 Egg are allowed.	2 oz. of cooked Meat. Chop. Light Pudding. Beef Tea (1 lb. of gravy beef to each pint). Eggs. Oysters. Green Vegetables. Bottled Ale and Stout.

UNIVERSITY COLLEGE HOSPITAL.

FULL DIET.
12 oz. Bread.
8 oz. Potatoes.
6 oz. Meat dressed (Roast or Boiled Leg or Neck of Mutton, or Roast Beef).
¾ pint of Broth or Pea-soup four times a week on alternate days.
4 oz. Boiled Rice or Rice Pudding made with Milk.
1 pint of Milk.
1 pint of Beer.*

MIDDLE DIET.
12 oz. of Bread.
8 oz. of Potatoes.
4 oz. Meat or 8 oz. of Fish (White).
1 pint of Milk.
Soup with Barley 1½ oz., or Beef Tea 1 pint.
Rice Pudding made with Milk, instead of Soup.
½ pint Beer.*

SPOON DIET.
2 pints of Milk.
1 pint of Beef Tea.
12 oz. Bread.
2 oz. Arrowroot and 1 oz. Sugar made into Jelly.

The following extras are supplied only on special orders:—Malt Liquors, Wines, Spirits, Eggs, Strong Beef Tea, Milk, Fish, Chops, Steaks, Custard Puddings, Vegetables, and Bread.

---

* To Medical Cases only when ordered.

WESTMINSTER HOSPITAL.

DIET TABLE.

Meals.	Full Diet.	Middle Diet.	Simple Diet.
Daily	14 oz. of Bread. ¾ oz. of Butter. ¼ oz. of Tea. 1 oz. of Sugar. ½ pint of Milk.	10 oz. of Bread. ¾ oz. of Butter. ¼ oz. of Tea. 1 oz. of Sugar. ½ pint of Milk.	8 oz. of Bread. ¼ oz. of Tea. 1 oz. of Sugar. ½ pint of Milk.
Breakfast	Tea and Bread and Butter.	Tea and Bread and Butter.	
Dinner	5 oz. of cooked Meat, roasted or boiled. ½ lb. of Potatoes.	3 oz. of cooked Meat, roasted or boiled. ½ lb. of Potatoes.	1 pint of Beef Tea. 1 pint of Milk.
Tea	Tea and Bread and Butter.	Tea and Bread and Butter.	
Supper	1 pint of Milk Gruel.	1 pint of Milk Gruel.	
Extras	Any of the following may be ordered in addition to, or in substitution of, the above Diets, viz.:— Extra Bread or Milk, Broth, Beef Tea, Ale, Porter, Barley Water, Soda Water, Lemonade, Lemons, Sugar, Green Vegetables, Butter, Rice or Bread Pudding, Arrowroot, Sago, Vermicelli, Eggs, Ice, Gruel. In substitution for Tea, Coffee or Cocoa may be ordered; and in substitution for other Meat, Fish, Chop, or Irish Stew.		

N.B.—Wines and Spirits are ordered only as Medicine.

WESTMINSTER HOSPITAL—*continued.*

Every Patient admitted into the Hospital is to be placed by the Resident Medical Officer on either Simple or Middle Diet, without any extra other than additional Beef Tea or Milk, until a Diet is ordered by the Physician or Surgeon.

No Extras other than those specified above are to be placed on the Diet Roll by the Steward, except on the special requisition of the Physicians and Surgeons.

All Extras are to be authorised by the signature of the Physicians and Surgeons, to be renewed twice in each week at least.

A return of all Extras ordered, together with the names of the Physicians and Surgeons ordering the same, is to be made weekly by the Steward.

QUANTITIES AND COMPOSITION OF THE ARTICLES OF DIET AS ABOVE.

Chop (uncooked) 5 oz.

Chop for Irish Stew, 5 oz. (with ½ lb. Potatoes, &c.).

Meat for pint of Beef Tea, 9 oz.

Potatoes, ½ lb.

Fish, 8 oz.

Butter, ¾ oz.

Milk Gruel to contain ¼ pint of Milk to the pint.

Coffee, 1 oz.
Cocoa, ½ oz.
Tea, ⅓ oz.
Sugar, 1 oz.

Bread Pudding—
  Bread, ¼ lb.
  Milk, ¼ pint.
  Sugar, ¼ oz.
  Flour, ¼ oz.
  1 Egg for every 2 lbs.

Rice Pudding—
  Rice, 1½ oz.
  Milk, ½ pint.
  Sugar, ½ oz.

HOSPITAL FOR WOMEN.

DIETS.

FULL DIET.
Daily 10 oz. Bread.
,, 1 pint of Milk.
,, ¾ oz. Butter.
,, 5 oz. Meat (cooked).
,, 1 pint of Beef Tea.
,, ¼ lb. Potatoes.
3 *Days* other Vegetables.

FISH DIET.
Daily 10 oz. Bread.
,, 1½ pint of Milk.
,, 1 pint of Beef Tea.
,, ¾ oz. Butter.
,, 6 oz. Fish.
,, ¼ lb. Potatoes.

MILK DIET.
Daily Bread *ad lib*.
,, 2 pints of Milk.
,, ¾ oz. of Butter.
,, 1 pint of Beef Tea.
,, ½ pint of Light Pudding.

EXTRAS.

Beer, Wine, Spirits, Mutton Chops, Beef Tea, Eggs, Chicken, Lemonade, Soda Water, Ice, &c.

No extras allowed unless ordered, *in writing*, by one of the Medical Officers.

N.B.—Patients are not allowed to receive from their friends any food, with the exception of tea, coffee, cocoa, and sugar.

BEEF TEA.

1 lb. of Rump Steak free from fat.
1. Mince fine.
2. Pour on half-a-pint of cold water, and stand three hours, stirring occasionally.

3. Pour off and press this and keep the fluid.
4. Add another half-a-pint of cold water to the same meat.
5. Stew but not boil for two hours.
6. Pour and press off fluid.
7. Add this to the fluid first reserved, and boil the two together.
8. Let it cool and skim off fat.

The sediment of this must be stirred and taken with the beef tea.

PHARMACOPŒIA

OF THE

GERMAN HOSPITAL,
DALSTON, near LONDON.

COLLYRIA.

Collyrium Aluminis.
: Aluminis Pur. 6 grs.; Tinct. Myrrh. 12 mins.; Aq. Destill. 1 oz.

Collyrium Argenti Nitratis.
: Argent. Nitrat. 2 grs.; Aq. Destill. 1 oz.

Collyrium Zinci.
: Zinc. Sulph. 2 grs.; Aq. Destill. 1 oz.

DECOCTUM.

Decoctum Cinchonæ.
: Decoct. Cinchon. Flav. P. B. 1 oz.: Acid. Sulph. Dilut. 6 mins.

ELECTUARIA.

Electuarium Sulphuris.
: Pulv. Sulphur. Co. ½ oz.; Syrup $q.\ s.$ (1 oz.).

Electuarium Vermifug.
: Pulv. Jalap. 16 grs.; Pulv. Valerian. 8 grs.; Pulv. Sem. Santonic. ½ drm.; Syrup $q.\ s.$ (1½ drm.).

EMULSIO.

Emulsio Camphor. c. Benz.
: Acid. Benzoic. 10 grs.; Camphoræ 2 grs.; Gum. Acaciæ ½ drm.; Aquæ ad 1 oz.

GARGARISMATA.

Gargarisma Aluminis.
> Aluminis 10 grs.; Aquæ 1 oz.

Gargarisma Potassæ Chloratis.
> Potass. Chlorat. 20 grs.; Aquæ 1 oz.

INJECTIONES.

Injectio Zinci.
> Zinc. Sulph. 3 grs.; Aq. Destill. 1 oz.

Injectio Zinci. c. Tannin.
> Zinc. Sulph. 3 grs.; Acid. Tannic. 3 grs.; Aq. Destill. 1 oz.

LINIMENTA.

Linimentum Camphoræ Co.
> Camphoræ 1 drm.; Sapon. Mollis 1 drm.; Liq. Ammon. Fort. 40 mins.; Spir. Vini ad 1 oz.

Linimentum Sulphuris.
> Sulphur. Præcip. 1 oz.; Spir. Vini Rect. 1 oz.; Glycerin. 1 oz.

LOTIONES.

Lotio Hispanic.
> Zinc. Sulph. 1 gr.; Plumb. Acetat. 1 gr.; Aq. Destill. 1 oz.

Lotio Sodæ. Chlorinatæ.
> Liq. Sod. Chlorinat. 6 mins.; Aq. Destill. 1 oz.

MISTURÆ.

Mistura Acaciæ.
> Mucilago Acaciæ 1½ drm.; Aquæ ad 1 oz.

Mistura Ammoniæ Acetatis.
> Liq. Ammon. Acet. 1 drm.; Aquæ ad 1 oz.

Mistura Ammoniæ Muriatis.
> Ammon. Mur. 30 grs.; Mucil. Acaciæ 1 drm.; Extr. Glycyrrh. Liquid. 30 mins.; Aq. Anisi ad 1 oz.

Mistura Antiblennorrh.
Bals. Copaibæ 30 mins.; Gum. Acaciæ 20 grs.; Aquæ ad 1 oz.

Mistura Anticatarrh.
Ammon. Mur. 12 grs.; Mag. Sulph. 30 grs.; Extr. Glycyrrh. Liquid. 30 grs.; Aquæ ad 1 oz.

Mistura Antidiarrhoic.
Tinct. Opii 8 mins.; Tinct. Catechu 30 mins.; Mist. Cretæ P. B. ad 1 oz.

Mistura Antiphlogist.
Mag. Sulph. 80 grs.; Tinct. Card. Co. 24 mins.; Aquæ ad 1 oz.

Mistura Colchici.
Vin. Colchici 24 mins.; Sol. Sodæ Bicarb. 1 oz.

Mistura Diuretica.
Potassæ Acetatis 24 grs.; Tinct. Scillæ 20 mins.; Spir. Ætheris Nitrosi 1 drm.; Infus. Digital. P. B. ad 1 oz.

Mistura Ferri Perchloridi.
Tinct. Ferr. Perchlor. 20 mins.; Infus. Quassiæ 1 oz.

Mistura Ferri Pomati.
Tinct. Ferri Pomat. 30 mins.; Infus. Quassiæ 1 oz.

Mistura Fowleri.
Liq. Potass. Arsen 10 mins.; Infus. Quassiæ 1 oz.

Mistura Morphiæ.
Liq. Ammon. Acet. 2 drms.; Spir. Ætheris 30 mins.: Liq. Morph. P. B. 20 mins.; Aquæ ad 1 oz.

Mistura Oleosa Gummosa.
Ol. Amygdal. Dulc. 30 mins.; Gum. Acaciæ 15 grs.; Aquæ ad 1 oz.

Mistura Stomachica.
Tinct. Zingiber. 6 mins.; Tinct. Cinchon. Co. 12 mins.; Tinct. Card. Co. 6 mins.; Infus. Quassiæ ad 1 oz.

Mistura Quiniæ c. Ferro.
M. Quin. Co. 1 oz.; M. Ferr. Perchlor. 1 oz.

PILULÆ.

Pilula Aloes c. Ferro.
Aloes Socotrin. 1½ gr.; Ferr. Sulph. Exsicc. 1 gr.: Extr. Gentian. 1 gr.: in each pill.

Pilula Doveri.
Pulv. Ipecac. Co. 3 grs.; Conf. Rosæ 1 gr.; Pulv. Glycyrrh. 1 gr.: in each pill.

Pilula Hydrarg. Bichl. c. Opio vel Antisyphilitic.
Hydrarg. Bichlor. Corros. $\frac{1}{16}$ gr.; Opii Pur. $\frac{1}{8}$ gr.; Conf. Rosæ 1 gr.; Pulv. Glycyrrh. 1 gr.: in each pill.

Pilula Hydrarg. Iodid. c. Opio.
Hydrarg. Iodid. Virid 1 gr.; Opii Pur. ¼ gr.; Conf. Rosæ 1 gr.; Pulv. Glycyrrh. 1 gr.: in each pill.

PULVERES.

Pulvis Adstringens.
Pulv. Cretæ Co. 4 grs.; Pulv. Ipecac. Co. ½ gr.; Pulv. Cinnam. Co. 2 grs.: in each powder.

Pulvis Alterans.
Pulv. Rhei 6 grs.; Mag. Carb. 4 grs.; Hydrarg. c. Cretâ 2 grs.; Pulv. Cinnam. Co. 2 grs.: in each powder.

Pulvis Calcis Phosphatis Co.
Calcis Phosphat., Calcis Carbonat., Sacch. Lactis, of each equal parts.

Pulvis Infant. Hufland.
Pulv. Rad. Iridis 90 grs.; Pulv. Rad. Rhei 120 grs.; Sacch. Alb. ½ oz.; Mag Carbon. 1 oz.; Ol. Fœnicul. 12 mins.

Pulvis Sulphur. Co.
> Sulph. Sublim. ½ oz.; Pulv. Fol. Sennæ 2 drms.;
> Mag. Carbon. 20 grs.; Potass. Bitart. 1 oz.;
> Pulv. Zingiber. 30 grs.

SOLUTIONES.

Solutio Ferri et Ammon. Citratis.
> Ferr. et Ammon. Citr. 10 grs.; Aquæ ad 1 oz.

Solutio Lugold.
> Pot. Iodid. 40 grs.; Iod. Pur. 20 grs.; Aquæ 1 oz.

Solutio Magnesiæ Sulphatis.
> Mag. Sulph. 80 grs.; Aquæ ad 1 oz.

Solutio Magnesiæ Sulphatis Alcalina.
> Sod. Bicarb. 12 grs.; Mag Sulph. 80 grs.; Aquæ ad 1 oz.

Solutio Plencki.
> Hydrarg. Bichlor. Corros. 1 drm.; Camphor. 30 grs.; Spir. Vini Rect. 1 oz.

Solutio vel Gargar. Potassæ Chloratis.
> Potass. Chlorat. 20 grs.; Aquæ 1 oz.

Solutio Potassii Iodidi.
> Potass. Iodid. 10 grs.; Aquæ Destill. 1 oz.

Solutio Quiniæ Co.
> Quiniæ Sulph. 2 grs.; Acid. Mur. Dilut. 4 mins.;
> Tinct. Aurant. 12 mins.; Aquæ ad 1 oz.

Solutio Sodæ Bicarbonatis.
> Sod. Bicarb. 24 grs.; Infus. Quassiæ 1 oz.

Solutio Sodæ pro Infante.
> Sod. Bicarb. 6 grs.; Syrup. 60 mins.: Aq. Destill. ad 1 oz.

Solutio Sodæ Sedativa.
> Aq. Lauro-Ceras. 30 mins.; Sol. Sodæ Bicarb. ad 1 oz.

Solutio Sodæ Sulphatis.
> Sod. Sulph. Cryst. 80 grs.; Aquæ ad 1 oz.

TINCTURÆ.

Tinctura Ferri c Morphiâ.
> Tinct. Ferr. Perchlor. 2 drms.; Liq. Morph. P. B. 1 drm.

Tinctura Ferri Pomati.
> Extr. Ferr. Pomat 60 grs.; Aq. Cinnam. Vinos ad 1 oz.

UNGUENTA.

Unguentum Arlti.
> Extr. Belladonn. 8 grs.; Hydrarg. Ammon. 6 grs.; Axungiæ 1 oz.

Unguentum Calomel. Co.
> Calomel. 60 grs.; Zinc. Sulph. 60 grs.; Axungiæ ad 1 oz.

Unguentum Calomel. c. Zinco.
> Calomel. 160 grs.; Zinc. Oxyd. 24 grs.; Axungiæ 1 oz.

Unguentum Hebræ.
> Empl. Plumbi 1 oz.; Ol. Lini 2 oz.; Axungiæ ½ oz.

Unguentum Sulphuris Co.
> Sulph. Sublim. 2 drms.; Sapon Mollis 1 oz.

VINUM.

Vinum Aromatica.
> Fol. Menth. Pip. 2 drms.; Flor. Anthem. 2 drms.; Caryophyll. 15 grs.; Cocc. Cacti. 10 grs., Sherry Wine 10 oz.: macerate seven days, strain, and add Acid. Tannic. 30 grs.; filter.

EDINBURGH.

The Chief Hospitals use the British Pharmacopœia.

PHARMACOPŒIA

OF THE

MEATH HOSPITAL AND COUNTY DUBLIN INFIRMARY.

COLLYRIUM.

Collyrium Zinci.
Acetate of Zinc 1 gr.; water to 1 oz.

HAUSTUS.

Haustus Olei Ricini.
Castor Oil ½ oz.; Solution of Potash 25 mins.; Tincture of Rhubarb 2 drms.; Cinnamon Water to 1½ oz. Half the dose for a child.

Haustus Rhei.
Powdered Rhubarb 20 grs.; Sulphate of Magnesia 60 grs.; Peppermint Water 1½ oz.

LOTIONES.

Lotio Calcis Chloratæ.
Solution of Chlorinated Lime 15 mins.; water to 1 oz.

Lotio Stimulans.
Comp. Tincture of Benzoin 15 mins.; Solution of Chlorinated Lime 15 mins.; water to 1 oz.

MISTURÆ.

Mistura Amara Acida.
Dil. Nitro-hydrochloric Acid 15 mins.; Infusion of Quassia to 1 oz.

Mistura Amara Alkalina.
Bicarbonate of Soda 11 grs.; Infusion of Quassia 1 oz.

Mistura Anti-Dyspeptica.
Cherry-Laurel Water 11 mins.; Solution of Lime 3 drms.; Essence of Peppermint 7½ mins.; Syrup of Ginger 22½ mins.; Comp. Tincture of Lavender 15 mins.; Infusion of Quassia to 1 oz.

Mistura Astringens Composita.
Tincture of Opium 11 mins.; Tincture of Catechu 15 mins.; Tincture of Kino 15 mins.; Chalk Mixture to 1 oz. Dose ½ oz.

Mistura Cardiaca.
Spirit of Ether 15 mins.; Aromatic Spirit of Ammonia 15 mins.; Comp. Tincture of Lavender 15 mins.; Camphor Water to 1 oz. Dose ½ oz.

Mistura Copaiba.
Copaiba 22½ mins.; Solution of Potash 7½ mins.; Tincture of Henbane 30 mins.; Spirit of Nitrous Ether 15 mins.; Cinnamon Water to 1 oz. Dose ½ oz.

Mistura Diaphoretica.
Solution of Acetate of Ammonia 2 drms.; Nitrate of Potash 5 grs.; Spirit of Nitrous Ether 15 mins.; water to 1 oz.

Mistura Diuretica.
Spirit of Juniper 30 mins.; Spirit of Nitrous Ether 15 mins.; Decoction of Broom to 1 oz.

Mistura Expectorans.
Wine of Ipecacuanha 11 mins.; Syrup of Squill 15 mins.; water to 1 oz. Dose 2 drms.

Mistura Pectoralis Sedativa.
Cherry-Laurel Water 11 mins.; Solution of Morphia 7½ mins.; Comp. Tincture of Camphor 15 mins.; water to 1 oz. Dose ½ oz.

Mistura Potassii Iodidi.
Iodide of Potassium 5 grs.; Comp. Tincture of Lavender 11 mins.; water to 1 oz.

Mistura Purgans.
Sulphate of Magnesia 109 grs.; Tincture of Ginger 15 mins.; Infusion of Senna to 1 oz.

Mistura Terebinthinæ.
Oil of Turpentine 7½ mins.; Mucilage 1 drm.; Spirit of Nitrous Ether 15 mins.; water to 1 oz.

Mistura Tonica et Aperiens.
Sulphate of Magnesia 54½ grs.; Tincture of Ginger 15 mins.; Infusion of Quassia to 1 oz.

PILULA.
Pilula Purgantes.
Colocynth and Hyoscyamus Pill 5 grs. in each.

UNGUENTUM.
Parr's Ointment.
Nitrate of Mercury Ointment, Sulphur Ointment, Tar Ointment, of each equal parts.

PHARMACOPŒIA

OF THE

ADELAIDE HOSPITAL, DUBLIN.

LINIMENTUM.

Linimentum Stimulans.
> White Soap 2 oz.; hot water 35 oz.; dissolve, and when cold add Strong Solution of Ammonia 5 oz.; Oil of Turpentine 10 oz.

LOTIONES.

Lotio Acidi Carbolici.
> Carbolic Acid 1 oz.; water 30 oz.

Lotio Calcis Chloratæ.
> Solution of Chlorinated Lime (B.P.) 2 drms.; water to 8 oz.

Lotio Nigra.
> Calomel 100 grs.; Lime Water 20 oz.

Lotio Plumbi.
> Solution of Subacetate of Lead (B.P.) 1 drm.; water 8 oz.

MISTURÆ.

Mistura Amara.
> Infusion of Quassia (B.P.) Dose ½ oz. to 1 oz.

Mistura Amara Aperiens.
> Infusion of Quassia 1 oz.; Sulphate of Magnesia 32 grs.; Diluted Sulphuric Acid 15 mins. Dose ½ oz. to 1 oz.

Mistura Amara c. Ferro.
   Infusion of Quassia 1 oz.; Sulphate of Iron 2 rs.
   Dose ½ oz. to 1. oz.

Mistura Arsenicalis c. Ferro.
   Hydrochloric Solution of Arsenic (B.P.) 10 mins.;
   Diluted Solution of Perchloride of Iron 10 mins.;
   water to 1 oz. Dose ½ oz. to 1 oz.

Mistura Cardiaca.
   Carbonate of Ammonia 4½ grs.; Camphor Water 1
   oz. Dose ½ oz. to 1 oz.

Mistura Carminativa.
   Tincture of Rhubarb 48 mins.; Carbonate of Magnesia 9 grs.; Aromatic Spirit of Ammonia 9
   mins.; water to 1 oz. Dose ½ oz. to 1 oz.

Mistura Copaibæ.
   Copaiba 15 mins.; Solution of Potash 9 mins.;
   Nitrate of Potash 9 grs.; water to 1 oz. Dose
   ½ oz. to 1 oz.

Mistura Cretæ Composita.
   Tincture of Catechu 12 mins.; Tincture of Rhatany
   12 mins.; Chalk Mixture (B.P.) to 1 oz. Dose
   ½ oz. to 1 oz.

Mistura Diaphoretica.
   Solution of Acetate of Ammonia 1 drm.; Nitrate of
   Potash 20 grs.; water to 1 oz. Dose ½ oz. to
   1 oz.

Mistura Diuretica.
   Acetate of Potash 20 grs.; Sweet Spirit of Nitre 30
   mins.; Decoction of Broom 1 oz. Dose ½ oz.
   to 1 oz.

Mistura Hydrargyri.
   Corrosive Sublimate ⅛ gr.; Sal Ammoniac ¼ gr.;
   Tincture of Cardamoms 30 mins.; water to 1 oz.
   Dose ½ oz.

Mistura Pectoralis.
   Carbonate of Ammonia 3 grs.; Tincture of Squill
   20 mins.; Syrup ½ drm.; water to ½ oz. Dose
   ½ oz. to 1 oz.

Mistura Pectoralis Infantilis.
Sweet Spirit of Nitre 15 mins.; Ipecacuanha Wine 7 mins.; Syrup ½ drm.; water to ½ oz. Dose 1 drm. to ½ oz.

Mistura Potassii Bromidi.
Bromide of Potassium 10 grs.: water 1 oz. Dose ½ oz. to 1 oz.

Mistura Potassii Iodidi.
Iodide of Potassium 10 grs.; water 1 oz. Dose ½ oz. to 1 oz.

Mistura Purgans.
Sulphate of Magnesia 2 drms.; Tincture of Senna 1 drm.; Ginger 7 grs.; water to 1 oz. Dose ½ oz. to 1 oz.

Mistura Sedativa.
Muriate of Morphia $\frac{1}{16}$ gr.; Diluted Hydrocyanic Acid 4 mins.; Syrup 1 drm.; water to 1 oz. Doze ½ oz.

Mistura Stomachica.
Bicarbonate of Soda 10 grs.; Ginger 5 grs.; Infusion of Quassia 1 oz. Dose ½ oz. to 1 oz.

Mistura Strychniæ.
Solution of Strychnia (B.P.) 5 mins.; Solution of Perchloride of Iron 5 mins.; water to 1 oz. Dose ½ oz. to 1 oz.

Mistura Terebinthinæ.
Confection of Turpentine 40 grs.; Peppermint Water to 1 oz. Dose ½ oz. to 1 oz.

PILULA.

Pilula Purgans.
Compound Gamboge Pill 3 grs.; Blue Pill 2 grs. Dose 5 to 10 grs.

PULVERES.

Pulvis Alterativus.
Grey Powder 1 gr.; Rhubarb 2 grs.; Bicarbonate of Soda 2 grs. Dose 5 grs.

Pulvis Ferri.
Saccharated Carbonate of Iron 6 grs.; Reduced Iron 1 gr. Dose 7 grs.

Pulvis Santonini.
Santonine 2 grs.; Compound Jalap Powder 5 grs. Dose 7 grs.

SKIN DISPENSARY.

Sapo Acidi Carbolici.

Tinctura Saponis Composita.
Soft Soap, Oil of Cade, Methylated Spirit, of each 6 oz.

Unguentum Acidi Carbolici.
Carbolic Acid 1 drm.; Lard 1 oz.

THE END.

www.ingramcontent.com/pod-product-compliance
Lightning Source LLC
Chambersburg PA
CBHW021209230426
43667CB00006B/627